The Garden of Ediacara

•

FRONTISPIECE: The Nama Group, Aus, Namibia, August 9, 1993. From left to right, A. Seilacher, E. Seilacher, P. Seilacher, M. McMenamin, H. Luginsland, and F. Pflüger. Photograph by C. K. Brain.

The Garden of Ediacara

•

Discovering the First Complex Life

Mark A. S. McMenamin

Columbia University Press
New York

Columbia University Press
Publishers Since 1893
New York   Chichester, West Sussex

Copyright © 1998 Columbia University Press

Library of Congress Cataloging-in-Publication Data
McMenamin, Mark A.
    The garden of Ediacara : discovering the first complex life / Mark A. S.
    McMenamin.
        p.   cm.
    Includes bibliographical references and index.
    ISBN 978-0-231-10559-0 (alk. paper)
    1. Paleontology—Precambrian. 2. Fossils. I. Title.
QE724.M364   1998
560'.171—dc21
                                                              97-38073
                                                              CIP

♾

Casebound editions of Columbia University Press books are printed on
permanent and durable acid-free paper.
Printed in the United States of America

For Gene Foley
Desert Rat par excellence
and to the memory of
Professor Gonzalo Vidal

Contents

Foreword

Dorion Sagan

Virtually as soon as earth's crust cools enough to be hospitable to life, we find evidence of life on its surface. But we are latecomers, and just as we must be familiar with the beginning of a mystery novel to understand its end, we must scrutinize the often ignored early phase of evolution. Mark McMenamin's allusively named *Garden of Ediacara* hones in on some of the key events and players in life's early phase—a time for the biosphere that, like the first three years of a human life, is not only formative and revealing but essential to understanding the full sweep of a living existence.

Da Vinci found shells on mountains that suggested a long geological past. Hutton and, later, Darwin extended such thinking, drawing forth a temporal expanse wide enough to explain modern anomalies and complexities. But when early commentators surveyed the fossil history of life on earth, they were not overly impressed with life's earliest phase. It almost seemed as if nothing was going on. Until the "Cambrian explosion"—the widespread appearance of fossil forms, including the famous horseshoe-crab-like trilobites, during the Cambrian geological period—it seemed as if life had barely started. Now you don't see them, now you do: Like the goddess Minerva bursting forth fully formed from the head of Zeus, the sudden appearance of hard-backed animals in the fossil record had about it the lingering aura of myth or celestial-fostered miracle.

Whence come animals from evolutionary chaos?

For geologist Preston Cloud, one of the first of the modern paleobiologists, the appearance of animal life corresponded to a global atmospheric increase in free oxygen. This theory, repeated in textbooks, may be an anthropomorphic fairy tale, a kind of industrial fiction. Fire-starting oxygen, the gas of choice, spurs the biosphere to produce complex life forms, paving the way for air-breathing mammals. But there is probably no causal relationship between oxygen increase and animal life. The Cambrian explosion was 540 million years ago, whereas according to the rock record of oxygen-rich uranium and iron ores, atmospheric oxygen began to build up far earlier, some 1800 million years ago. Nothing is so destructive of a beautiful theory as an ugly fact.

Although classical evolutionists pictured a gradual evolution of animal life from soft-bodied to hard-bodied forms, the shelled creatures of the Cambrian stick out like a sore thumb. There does seem to be a suddenness about them, one not explicable on the basis of gradual evolution.

Today we understand that the Cambrian fauna were preceded by a strange and motley collection of often symmetrical soft-bodied forms. These are the Ediacarans, eponymous subheroes of the Australian outcrop where the first such fossils were found. The Ediacarans' global "garden," more than a cryptic play on Eden's idyllic and instantaneous fertility, refers to their largely vegetative existence. With the playful attitude of a true scientific explorer, Mark McMenamin treks the planet and mines the literature, some of it itself almost fossiliferous, in an exposition of medusoids, ring stones, concentrically fretted, radially flaring, and other enigmatic trace and body fossils left by the soft-bodied pre-Cambrian forms. Who were these beings? Were they animals? Our ancestors?

More likely, we find, they were our cousins. Although superficially similar to jellyfish, the Ediacaran medusoids probably never swam: They are preserved concave side up, like bowls rather than like swimmers. They may have been so quickly replaced all over the world not as a result of evolution-igniting oxygen, but because evolution had gotten to the point where predators with eyes and a murderous appetite for Ediacaran sushi had come into their own. In McMenamin's persuasive reading, the earliest animals (or animal-like life, for they may have been blastula-less colonial microbes called protoctists—predating predation) were languid, limpid vegetarians. Harmless antecedents to Tennyson's bloody nature tooth and claw, they were eyeless representatives of a victimless Edenlike world. This was a green and serene world where there was no reason for calcified coverings, for carapaces or spiky armor because the victimizing element of Animalia had not yet evolved.

The Ediacarans, on this view, were translucent beings with photosynthetic inclusions, soaking up the sun and living off the excess of their living internal gardens. Today creatures such as the snail *Placobranchus*, the giant clam *Tridacna*, and the seaweed-looking worm *Convoluta roscoffensis*, whose mouth is closed throughout its adult life, have gone back to the simpler "Edenic" lifestyle of autotrophic sunbathing. Ryan

Drum has even made the semiserious suggestion that this would be a good thing for junkies: Inject them with algae, select for ever more pallid and translucent demeanor, and in time we might have societally harmless anthropoids nutrifying themselves in languor at a delicious if safe remove from the normal frenetic hustle of urban animal life. The photosynthetic cells might even migrate to the germ cells of these vegetablarians, permitting true speciation of *Homo photosyntheticus* from *Homo sapiens* ancestors.

It would be a mistake to think such "planimals" only freaks of symbiosis or figments of science fiction. All plants and algae on the planet are composed of eukaryotic cells whose existence has been brought to us by primeval mergers with cyanobacteria, green prokaryotes that were eaten (living salad) by larger cells. Blessed with a permanent case of indigestion, these larger cells benefitted from the metabolic independence of the preplants now dwelling Jonah-like inside them. McMenamin's Garden of Ediacara hypothesis explains the widespread soft-bodied beings, many flat and fronded, as early testimony to the power of photosymbiosis. By turns starfishy and Star Treky, branched and sand dollaresque—switching, like some fossil equivalent to a Necker cube, from twisting worm to growing stem in the paleontologist's imagination— these creatures may well have been neither plant nor animal. On the television program *The X Files*, FBI agents Sculley and Mulder find "a cell that is not a plant cell or an animal cell. And it's dividing mitotically." How creepy! How strange! The only problem with this would-be biological bizarrerie is that such uncanny cells are more prevalent on the surface of earth than either plant or animal cells. They're in your house right now, and they live in your body. Bacteria, although they do not divide mitotically, are neither animals nor plants. And their symbiogenetic offspring, the protoctists, which do divide mitotically (and which are on your skin right now), were certainly among the ancestors to the Ediacarans. But the "metacellularity" of the Ediacaran organisms may not have led to any large extant forms of life familiar to us today. They were jellyfish-*like* and starfish-*like*, but neither truly medusoids nor echinoderms. Some secreted copious mucous layers as a means of gliding locomotion. True aliens from our own past, some of these fabulously real beings lost their innocence and symmetry. Concentrating sense organs at one end, some may have been, before animals proper, the first organisms to evolve heads.

Which, as in the knowledge gained from the Tree of Good and Evil in the redolent Garden of Eden, may have spelled the beginning of the end.

For despite the scene drawn by some evolutionists of a planet of pure contingency, devoid of direction, certain patterns do have a way of cropping up again and again in evolution. Eusocial animals, for example, come not only in bee and ant but in naked mole rat flavors. Humans too, losing the inevitable link between sex and reproduction, may be moving toward eusociality. Like a tale told by a stutterer, evolution doubtless is circuitous and may well contain only the raw material, rather than the denouement, of meaning. But there are these patterns. Not just hominids but many species of mammal, for example, show an increase in body and brain size over evolutionary time. And the concentration of a suite of sense organs (some, perhaps, like the magnetodetecting abilities of some bacteria, alien to us) in the head, McMenamin intuits, may well have been an Ediacaran foreshadowing of the headstrong human theme. The consequences of this, as of Eve's bite, were severe. With eyes animal predecessors began to sense who was tasty, who was vulnerable—and to eat them. The evil empire of carnophagy had begun.

Henceforth, in this appealing and evidence-backed story, organisms quickly perished from lack of protection. Sashimi, it seemed, was everywhere. The arms races of predators and prey, the coevolutionary gamesmanship of becoming faster, smarter, and more deadly, on the one hand, and still faster, outsmarting, and quicker to hide or get away, on the other, came into being. The continents shifted. Ediacarans suffered. The innocence, or at least the languor, of the primeval Garden was lost.

Like all truly interesting stories, McMenamin's is a remarkable combination of speculation and fact. The virtue of the book you are about to read is that it enchants. Indeed, I bet Jorge Borges would have included an Ediacaran had he read McMenamin before scripting his *Book of Imaginary Beings*. The main difference between Borges's work and McMenamin's of course is that the present papers, "dubiofossils" notwithstanding, contain pictures and records of mainly real beings. Here, then, we have a work with all the allure of a medieval bestiary, with the difference that the creatures herein derive their mystery not so much from a distant, make-believe place as from long-elapsed time. It is a testament to McMenamin's success that he re-presents the Ediacaran garden, fleshing out a hypothesis and making real a time before life had even got its first bones. Take your ticket and get on board.

Preface

This book is the result of an ongoing and sometimes heated discussion among scientists with shared interests in the origins of animals. The conversation takes place on seacoasts, in deserts, in classrooms, in Newfoundland coffee shops over cod cheeks, at conferences and in symposia, and now with lightning speed over e-mail. I am pleased to present what I consider to be the definitive solutions to several vexing paleontological problems involving an unusual group of fossil organisms called the Ediacarans. In my view, the solutions to the Ediacaran problems are of utmost importance for our understanding of the world and the life it contains. However, I have not given the last word on these matters. The conversation will continue.

Mark McMenamin
South Hadley, Massachusetts

Acknowledgments

The author gratefully acknowledges the assistance of C. K. Brain, D. Evans, A. Fischer, S. Fossel, C. Franklin, W. Frucht, P. F. Hoffman, J. Hurtado, M. Johnson, J. Kirschvink, W. Krumbein, A. MacEachran, L. Margulis, D. L. S. McMenamin, P. Nevraumont, F. Pflüger, R. Riendeau, S. Rowland, D. Schwartzman, A. Seilacher, J. Stewart, and B. Stinchcomb. Memo to Connie Barlow: You were the first person to suggest I write a book with this title, so here it is. Funding for this research was provided in part by the National Science Foundation.

1 · Mystery Fossil

Your whole creation is never silent and never ceases to praise you. The spirit of every man utters its praises in words directed to you; animals and material bodies praise you through the mouth of those who meditate upon them, so that our soul may rise out of its weariness toward you, supporting itself upon the things which you created, and then passing on to you yourself who made them marvelously.

—St. Augustine[1]

[At] the dawn of European civilization, with the Greek philosophers, there were two clear tendencies in this problem. Those are the Platonic and the Democritian trends, either the view that dead matter was made alive by some spiritual principle or the assumption of a spontaneous generation from that matter, from dead or inert matter.

The Platonic view has predominated for centuries and, in fact, still continues to exist in the views of vitalists and neovitalists.

The Democritian line was pushed in the background and came into full force only in the seventeenth century in the work of Descartes. Both points of view really differed only in their interpretation of origin, but both of them equally assumed the possibility of spontaneous generation.

—A. I. Oparin[2]

Until not so long ago we thought that man had been specially created and that maggots arose from rotten cheese by spontaneous generation. It didn't matter, but now we believe that human beings have been evolved and it matters a very great deal. Thus, it is of the utmost importance that we should get to the truth of this matter.

—J. B. S. Haldane, introducing Oparin[3]

Coming across an arresting full-page illustration in the colorful *Time-Life Nature Library*, I became aware for the first time of the appeal of Ediacaran organisms. The illustration (figure 1.1), in stark black and white, showed an odd disc-shaped fossil, fringed by fine radial lines, with three curving arms at its center. The picture was the frontispiece for chapter 2, "The Origin of the Sea," and its caption was as telegraphic as a personals ad:

> MYSTERY FOSSIL, first of its rare kind ever found, has no known relationship with any other creature living or dead. It is also one of the oldest ever found. It comes from Pre-Cambrian rock strata in South Australia.[4]

This image held my attention and years later, when I had an opportunity to study Precambrian fossils at the University of California at Santa Barbara, I already appreciated the appeal of Ediacaran paleontology. In fact, I embarked on a study of the Ediacaran fossils for my postgraduate work.

These fossils are still as mysterious as when *Tribrachidium* was illustrated by the *Time-Life Nature Library* in the 1960s. With the Ediacaran fossils, or Ediacarans, paleontologists work a complex interface between the knowable (but difficult to know) and the unknowable (and thus outside the realm of science). The fossils of Ediacara document the events leading up to most important event in the history of life on earth.

FIGURE I.I: *Tribrachidium.*

Life has been part of this 4.45-billion-year-old planet for more than 3.5 billion years; calling Earth a living planet is more than just a poetic image. Life is now seen as both a process (a verb, in the view of Lynn Margulis and Dorion Sagan[5]) and an important geophysical and geochemical phenomenon. This sentiment was nicely stated by A. I. Oparin in 1965:

> As a rule the attempt to discover the possibility of life on Mars, Venus and other places has been made by the following methods. Studies were made of the conditions prevailing on these planets, and the question was asked if under these conditions organisms resembling those on Earth could exist. This is a fallacious approach. Life is produced by a certain environment, and it changes and alters the environment to adapt itself to it and adapt the environment to itself.[6]

Oparin's words seem strikingly current in light of the recent interest in the possibility of life on Mars. We know two things about life's origin. First, as Oparin pointed out in 1924, life originated in the absence of life and in the absence of free oxygen. Second, the appearance of life on Earth was apparently not a lengthy process. The earliest bacteria appeared almost as soon as Earth's crust was cool enough to support life. Oparin felt that "a billion years are needed to realize"[7] life's origin from inorganic precursors, but the geological record does not allow this much time for what might have been the first event of spontaneous generation. If the recent claims of ancient Martian life are true, then either life planet-skipped by some sort of phenomenon of panspermia or life is very easy to create under the proper physical and chemical conditions.

In our solar system at least, life could not get a foothold on a planet until the megacratering crisis had ended. This crisis was the period of early bombardment of planets by planetesimals (gigantic, subplanetary-sized meteors). An accretion of meteors such as these formed the planet in the first place. So much energy was released with each incoming rocky mass that the entire planetary surface was melted and presumably sterilized. This era of meltdowns has been called the "impact frustration of life" and ended on earth with the end of the intensive period of megacratering (as indicated by the ages of craters on the moon) about 3.8 billion years ago.

Rocks struck by meteoric impacts become pervasively fractured. When these fractures became fluid-filled, their surface areas expanded greatly, and thus may have become ideal sites,[8] precisely the microchemical factories needed for the origin of life. From a biological point of view, the fractures formed by incoming meteors represented a megacratering opportunity rather than a crisis.

The earliest life must have been microbial, the first forms probably being about .005 millimeter in diameter. A fascinating question concerning the origin of life is, "When did the first cell acquire the ability to distinguish self from nonself?" As A. G. Cairns-Smith argued in *Seven Clues to the Origin of Life*, life's origin may well have been as much a mineralogical phenomenon as a biochemical phenomenon.[9] In his view, a crystalline form of life (Gene-1) gave rise to a fully "organic" form of life (Gene-2).

Cairns-Smith felt that there must have been some sort of inorganic scaffolding on which the earliest life would have started. He proposed clay as the living crystal of Gene-1. More recent research has shown that clay does not have the properties needed to act as the scaffolding of Gene-2. Nevertheless, the main biomolecular constituents of life (nucleic acids, proteins, and phospholipids) are the products of complex biochemical synthesis pathways that cannot have arisen, de novo, on their own. As with self-supporting stones in a stone archway, some sort of scaffolding must have supported the stones during construction.

The idea of earliest life lacking individualization, forming as something like a living crystal, an extended body form that permeated some special environment of Earth, is indeed attractive. But however the first cells came to be, life apparently remained unicellular for billions of years. Multicellular life, individuals composed of billions or trillions of cells, did not appear on the globe until long after life began.[10]

The earliest organisms thought to represent multicellular creatures are uninspiring as fossils, occurring as more or less shapeless organic films (carbonized impressions) on slabs of shale.[11] The best that can be said about them, and this is by no means certain for all examples, is that they were eukaryotes, bearers of nucleated cells.

Eukaryotic cells are characterized by the presence of intracellular organelles, many of which were once free-living, and subsequently symbiotic, bacteria. This idea of a symbiotic origin for the organelles of eukaryotic cells gained momentum in the United States with Oparin's attendance at a conference in Wakulla Springs, Florida, in 1963. In the discussion session, Oparin presented the revolutionary idea of symbiogenesis, the thought that a new type of organism can emerge by the fusion of two unrelated types.[12] This was the first time many of the Western conference participants had heard these ideas:

The American investigator Hans Ris, of Wisconsin, visited the Soviet Union and has advanced an idea similar to what was expounded several years ago in Russia by Mereshkovskii, namely, that a cell represents a symbiotic structure. They said that for the time being the idea was rather too audacious. But it is possible you could develop it in the direction of representing the formation of cells as a gradual association, aggregation of symbionts.

Hans Ris was Lynn Margulis's adviser in college; to my surprise, before I mentioned it to her in June 1996, she had never heard that he had visited Russia. There appears to be a fascinating and untold story about the development of symbiogenesis theory in Soviet Russia, a story that may have its share of Cold War intrigues. However, I am not surprised by Oparin's comments because he was one of the few scientists of his stature at the time to have had more than a passing familiarity with symbiogenesis theory. The only other was the great Russian geologist Vladimir Ivanovich Vernadsky (1863–1945), who studied under symbiogeneticist Andrei S. Famintsyn, "founder of the Russian school of plant physiology, who demonstrated the possibility of photosynthesis in artificial light."[13] Vernadsky used his knowledge of symbiogenesis to found the now burgeoning field of biogeochemistry. In Vernadsky's view, biological processes are so important for our planet that it may truly be said that "life makes geology."

As Douglas R. Weiner points out in his review of Liya Nikolaevna Khakhina's book *Concepts of Symbiogenesis: A Historical and Critical Study of the Research of Russian Botanists* (translated into English in 1992),[14,15] symbiogenesis is integral to the Russian traditions in the history of science. Andrei S. Famintsyn (descended from a sixteenth-century Scottish immigrant whose name is the Russian translation of Thompson)[16] sought to supplement Darwinism with symbiogenesis theory. Konstantin S. Mereshkovskii tried to displace Darwinism with his new symbiogenesis theory between 1900 and 1920. Boris M. Kozo-Polyanskii tried to incorporate symbiogenesis smoothly within the overall schema of Darwinian evolution. Khakhina explains the slow headway symbiogenesis theory made in most scientific circles outside Russia. She describes it in terms of the perception that through the 1950s, symbiogenesis did not accord with the prevailing explanations of evolution.

The idea of a symbiotic origin of organelles is now the accepted theory presented in biology courses throughout the world. Nevertheless, scientists who espouse symbiogenesis raise hackles among their colleagues in evolutionary biology. One response to the murmuring is to boldly point out that there are indeed problems with the 1950s explanation of evolution, commonly called the neo-darwinian modern synthesis. A strong case can be made that neo-darwinism is due for an intellectual shakeup, and we return to this debate in chapter 13. As we will see, the solutions to the mysteries of Ediacara will play an important role in updating the modern synthesis. We start at the beginning of the Ediacaran fossil record.

The first large, complex, unquestionably multicellular fossils appear about 600 million years ago in stratified rocks of northern Mexico (chapter 9). Complex life on land, recognized by my wife Dianna and me as the biogeophysical entity Hypersea, appears some 200 million years later.[17]

Hypersea is the sum of eukaryotic life on land and all its symbionts. Despite its geological youth, Hypersea overwhelms the marine biota in terms of both total biomass and total biodiversity. This happens because the fluid connections between eukaryotes on land (particularly the ones involving plants and their root or mycorrhizal fungi) lead to a pumping of nutrients from the soil up into the photosynthetic parts of plants, a phenomenon we call hypermarine upwelling. Oparin[18] neatly anticipated our Hypersea theory, even hinting at hypermarine upwelling back in 1963: "Imagine that land life did not exist. From the standpoint of a jellyfish, life on dry land is sheer nonsense. Through a complex process of adaptation, of water exchange of circulation [sic], such a form of life was able to arise."

The first complex multicellulars and Hypersea are separated by the great divide in the geological time chart, the Precambrian-Cambrian boundary. This boundary is marked by what has been called the Cambrian breakthrough, the abrupt appearance of virtually all major types of skeleton-bearing animals. A robust and continuing evolutionary debate regarding this breakthrough[19] involves two main questions. First, did all the skeletalized animals appear suddenly at this time (the bang hypothesis), or do they have long histories that happened to leave virtually no fossil record (the whimper hypothesis)? Some authors advocate the whimper,[20] others the bang.[21] The whimperers are forced to admit that there is a major evolutionary radiation at the beginning of the Cambrian, although they try to keep the perceived number of new phyla appearing at this time to a minimum. The bangers see the phyla developing rapidly, and some postulate an unusual genetic reorganization that happens only at this time and is frozen into place (the green genes hypothesis, a version of the bang hypothesis).[22] The main problem with this putative fixing of particular gene expressions is that it is difficult or impossible to test scientifically.

The main proponent of the green genes hypothesis, James W. Valentine of the University of California at Berkeley, does not support the more extreme statements of his idea, and says that the elaboration of early animal genes "may have been necessary, but . . . was not sufficient, to drive the evolutionary creativity of the Cambrian."[23] His 25-year quest to explain the Cambrian explosion in terms of gene regulation has not yet met with unequivocal success.[24] Each successive Valentine paper on this subject seems to say, "Here is the latest breakthrough in modern genetic research; it must have something to do with the Cambrian explosion!" However, I believe that the origin of the major gene complexes in animals, an interesting subject in itself, has no necessary connection to the Cambrian event, and in fact may have been completely decoupled from it, the major steps in the formation of the animal genetic code having been taken well before the Cambrian.[25] There will be a better harvest for scientists among fossils and the ecological issues of the Garden of Ediacara.[26]

Bang or whimper, the Cambrian armored animals include many of familiar types that can be placed in still extant phyla. But for at least 50 million years before the Cambrian explosion, there existed a marine world of large[27] and unusual creatures.

These organisms constitute the Ediacaran biota. They have also been called the Ediacaran fauna, but because the term *fauna* implies animals, and paleontologists are not confident that all of the Ediacaran forms were animals; prudence requires the less specific term *Ediacaran biota*, or simply *Ediacarans*.

Diverse communities of multicellular creatures appear with the first members of the Ediacaran biota. My recent find in Mexico of trace fossils associated with the oldest Ediacarans indicates that true animals were unquestionably part of the biota.[28] Also present were the Ediacaran body fossil forms, less easily classified.

The Ediacaran biota seems at first glance to be another case of apparent spontaneous generation. Oparin's billion years are not evident here. My field research indicates that the Ediacarans sprang forth, fully formed, without a long record of evolution. This leads to the second question.

How could this happen? Furthermore, what kind of creatures are represented by the Ediacarans? Were they the first animals? They certainly

FIGURE 1.2: Seilacher's interpretation of the structure of Ediacarans. Left: Inflated, as in life. Right: Deflated, as in many fossil specimens. Note the rigid vertical walls. From M. A. S. and D. L. S. McMenamin, *The Emergence of Animals: The Cambrian Breakthrough* (New York: Columbia University Press, 1990). Artwork by Dianna McMenamin.

seem to be associated with trace fossil evidence of the earliest animals, but in the view of German invertebrate paleontologist Adolf Seilacher, they are not animals at all. In 1983 Seilacher destabilized what had been the consensus viewpoint (that is, Ediacarans as early animals) by pointing out that they had a quilted body architecture (figure 1.2) totally unlike anything seen in animals. Following insights made by German paleobotanist Hans D. Pflug, Seilacher argued that Ediacaran forms were *sui generis*, representatives of a group of high taxonomic rank[29] that went extinct at the beginning of the Cambrian.

A well-known science writer, following Seilacher's story, called the Ediacaran forms "aliens here on earth," meaning that they represented an alien body form no longer represented in the world.[30] Later work has demonstrated that these forms survived well into the Cambrian. However, the newer research has not settled the question of what these forms were, or how they fed. Many mysteries remain. The solutions may well involve a fuller understanding of the phenomenon of symbiogenesis. The question of the origin of life is an enduring puzzle, but we are just as ignorant about the origin of complex life.

Notes

1. Book V:1, p. 90 in Saint Augustine of Hippo, *The Confessions of St. Augustine,* translated by R. Warner (New York: Mentor, 1963).

2. S. W. Fox, ed., *The Origins of Prebiological Systems and of Their Molecular Matrices* (New York: Academic Press, 1965).

3. S. W. Fox, 1965.

4. The illustration of the fossil *Tribrachidium heraldicum* appears in L. Engel, *The Sea* (New York: Time-Life Books, 1969), 36–37.

5. L. Margulis and D. Sagan, *What Is Life?* (New York: Simon & Schuster, 1995).

6. See pp. 91–92 of A. I. Oparin, "History of the Subject Matter of the Conference," in S. W. Fox, ed., *The Origins of Prebiological Systems and of Their Molecular Matrices* (New York: Academic Press, 1965), 91–98.

7. See p. 345 in S. W. Fox, ed., *The Origins of Prebiological Systems and of Their Molecular Matrices* (New York: Academic Press, 1965).

8. Because of small-scale cation exchange and close association of clays, apatite, and other phosphate minerals.

9. A. G. Cairns-Smith, *Seven Clues to the Origin of Life* (Cambridge, England: Cambridge University Press, 1991).

10. At least a billion years after the origin of life.

11. H. J. Hofmann, "Paleocene #7 Precambrian Biostratigraphy," *Geoscience Canada* 14 (1987):135–154.

12. See p. 345 in S. W. Fox, ed., *The Origins of Prebiological Systems and of Their Molecular Matrices* (New York: Academic Press, 1965).

13. A. L. Yanshin and F. T. Yanshina, "The Scientific Heritage of Vladimir Vernadsky," *Impact of Science on Society* 151 (1988):283–296.

14. D. R. Weiner, "Book Reviews Feature Review," *Isis* 87 (1996):140–210.

15. L. N. Khakhina, *Concepts of Symbiogenesis: A Historical and Critical Study of the Research of Russian Botanists,* Lynn Margulis and Mark McMenamin, eds. (New Haven: Yale University Press, 1992).

16. M. B. Saffo, "Evolution of Symbiosis," *BioScience* 46 (1996):300–304.

17. C. Zimmer, "Hypersea Invasion," *Discover* 16, no. 10 (1995):76–87; M. A. S. McMenamin and D. L. S. McMenamin, *Hypersea: Life on Land* (New York: Columbia University Press, 1994).

18. See p. 92 in S. W. Fox, ed., *The Origins of Prebiological Systems and of Their Molecular Matrices* (New York: Academic Press, 1965).

19. L. M. Van Valen, "Review of *The Emergence of Animals: The Cambrian Breakthrough* by M. A. S. McMenamin and D. L. S. McMenamin, 1990, Columbia University Press," *Evolutionary Theory and Review* 10 (1992):172.

20. R. A. Fortey, D. E. G. Briggs, and M. A. Wills, "The Cambrian Evolutionary 'Explosion': Decoupling Cladogenesis from Morphological Disparity," *Biological Journal of the Linnaean Society* 57 (1996):13–33.

21. D. H. Erwin, J. W. Valentine, and D. Jablonski, "The Origin of Animal Body Plans," *American Scientist* 85 (1997):126–137. My favorite parts of this article are the illustrations; note the menacing gaze of the stalking anomalocarid on the cover illustration. Note also the use of Marilyn Monroe as representative of our branch of the animal family tree (M. DeRose, "Letters to the Editors." *American Scientist* 85 [1997]:204).

22. M. A. S. McMenamin and D. L. S. McMenamin, *The Emergence of Animals: The Cambrian Breakthrough* (New York: Columbia University Press, 1990).

23. See p. 137 of D. H. Erwin, J. W. Valentine, and D. Jablonski, "The Origin of Animal Body Plans," *American Scientist* 85 (1997):126–137.

24. See J. W. Valentine and C. A. Campbell, "Genetic Regulation and the Fossil Record," *American Scientist* 63 (1975):673–680; J. W. Valentine, "Late Precambrian Bilaterians: Grades and Clades," *Proceedings of the National Academy of Sciences USA* 91 (1994):6751–6757. See also B. Holmes, "When We Were Worms: The Garden of Ediacara," *New Scientist* 156 (1997): 30–35.

25. These steps were essentially complete by 600 million years ago, as shown

by the presence of trace fossils of this age in Mexico; see M. A. S. McMenamin, "Ediacaran Biota from Sonora, Mexico," *Proceedings of the National Academy of Sciences USA* 93 (1996):4990–4993.

26. M. A. S. McMenamin, "The Garden of Ediacara," *Palaios* 1 (1986):178–182.

27. Some up to a meter or more long.

28. Trace fossils are the markings made in sediment by the burrowing or locomotion of animals.

29. Such as a phylum or kingdom.

30. R. Lewin, "Alien Beings Here on Earth," *Science* 223 (1984):39.

2 · The Sand Menagerie

We often learn more from bold mistakes than from cautious equivocation.

-Daniel Dennett[1]

Aspiring paleontologists are typically attracted to the large, flashy specimens such as carnivorous dinosaurs and Pleistocene mammals. But to find the real monsters, the weird wonders of lost worlds, one must turn to invertebrate paleontology. Without question the strangest of all fossilized bodies are to be found among the Ediacarans.

The study of these forms, wrapped in mystery and founded on error, begins in 1856 with the publication of Ebenezer Emmons's description of what he called the "oldest organic bodies yet discovered."[2] Emmons, now famous for his discovery of the earliest Cambrian fossils known during his lifetime[3] and for his description of the first mammal-like reptile, was attacked in his day by his most distinguished contemporaries. However, no one questioned his description from Montgomery County, North Carolina, of the strange and very ancient new species *Palaeotrochis major* and *Palaeotrochis minor*. *Palaeotrochis*, or "old messenger," was described by Emmons as a disc- or spindle-shaped form covered on both sides with radial striae. Viewed from the top, the form resembles a medusoid, with radial striae emanating from a central boss.[4] Emmons described the new forms as fossil corals, thus beginning a tendency to view the earliest known or reputed animal fossils as members of the phylum *Cnidaria* (jellyfish, corals).

Emmons's report of the old messenger proved to be a faulty one; *Palaeotrochis* can now be shown to be a pseudofossil.[5] The fossils are associated with auriferous pyrite,[6] which accounts for the pseudofossil nature of Emmons's forms: They are pyrite rosettes. Emmons was the first but by no means the last geologist to interpret such forms as medusae of cnidarians.

Preston Cloud debunked some of the more recent of these in a 1973 paper titled "Pseudofossils: A Plea for Caution," although he was apparently unaware of Emmons's work on *Palaeotrochis*.[7] Ironically, true

Ediacaran fossils are now known to occur not far from the *Palaeotrochis* locality, in Stanly County, North Carolina.

Other messengers bear more reliable information about the past, and what follows is not meant to be an exhaustive catalog of all the different types of Ediacaran fossils known, but rather an introduction to all the main types of Ediacaran soft-bodied organisms. The few hard-bodied organisms of their time are dealt with in chapter 6.

It is impossible to organize these fossils into their systematic placement (correct taxonomic ordering), for there is no agreement on the biological affinities of any of these forms. This situation is a source of embarrassment for the science of paleontology. I attempt to rectify this situation later in this book, but for now, let us carefully examine the range of form of these curious fossils.

In this section I violate a conventional rule in science by deliberately mixing together discussions of observation and interpretation. It is usually prudent in formal scientific writing to carefully separate one's observations (the "results" section of the paper) from one's interpretations (the "conclusions" section of the paper). But my objective here, following in the footsteps of Seilacher, is to stir the pot, to inspire more creative thought about the Ediacarans. Also, it would be folly to pretend to completely separate one's observations from one's interpretations when writing about mysterious fossils. Any such pretenses would only obscure the important issues at hand. As a colleague once quipped, if we knew what we were doing, it wouldn't be research. So here we must begin.

Medusoids

The most common Ediacaran body fossils were the first type discovered at the classic site in the Ediacara Hills, South Australia. Circular or discoid Ediacarans are conventionally called medusoids, in an intentional comparison to the free-swimming medusa phase of the jellyfish life cycle. This is perhaps unfortunate because it is by no means certain that Ediacaran medusoids were related to the jellyfish medusa. However, the name has stuck.[8] The term *medusa* refers to the tentacles of living jellyfish, which resemble the snake-hair of Medusa, one of the three Gorgons slain by Perseus in Greek mythology.

In a dry and barren series of hillocks hundreds of kilometers north of Adelaide, Australia, R. C. Sprigg, who in 1946 was assistant government geologist of South Australia, was reassessing a series of abandoned lead-silver mines. The South Australian government was in the process of reviewing its mineral resources, and Sprigg had been sent to determine whether these mines were worth reopening.

While traversing the quartzite outcrops and flaggy slabs in the vicinity of the mines, Sprigg was pleasantly surprised to find unusual fossils in the quartzite exposures to the southwest of the mine area. Sprigg published several reports on the fossils and concluded that they all lacked hard parts and appeared to represent varied types of simple, ancient animals.[9] Sprigg went further to suggest that these fossils were some of the oldest evidence known of animal life.

Oddly enough, considering the importance of the find, more than a decade was to pass before a serious paleontological research expedition was mounted to the region. Led in October 1958 by Brian Daily, curator of paleontology at the South Australian Museum, the expedition returned with two trucks and a trailer full of fossils, a haul of over 1500 specimens.[10]

Sporadic collecting expeditions to this site continue to this day. Some specimens have made it onto the collectors' market; an acquaintance of mine recently purchased a small *Dickinsonia* specimen for $500 at a rock and mineral show in Springfield, Massachusetts. The specimen originally cost $600, but the vendor agreed to take $100 off the price if my colleague could identify the genus, which he proceeded to do.

Ediacarans from the type locality have also left Australia illegally.[11] In the early 1990s German fossil smugglers removed Ediacaran specimens from outcrops with motorized rock saws and shipped them to Asia. The fossils fell into the hands of Japanese collectors who paid for the specimens a sum reported to be in the high six figures in American dollars.

The thieves had committed an error as well as a crime, selecting a specimen that, from photographs taken on the outcrop, was well known to the paleontological community. Australian authorities notified Interpol, the specimens were seized in Japan, and the German smugglers were apprehended. The Japanese collectors found themselves shy a substantial sum of money.

The same smugglers were apparently active elsewhere in Australia and, in a raid of an important Early Cambrian trilobite locality, inadvertently exposed the first known occurrences of the Cambrian predator fossil *Anomalocaris* from the southern hemisphere. Fortunately for paleontologists, the smugglers did not recognize what they had found as they furtively dug for trilobites and tossed aside the anomalocarid specimens as if they were worthless matrix.

Medusoids are both the youngest and the oldest of the Ediacarans. They are also the most common, and were the first to be noted by scientists, although from the start there were questions about the biological nature of these structures. In 1877 E. Hill and T. G. Bonney reported "curious arrangements of concentric rings which have been supposed to be organisms,"[12] but then dismissed them as being accidental and inorganic.[13] These structures had been known to local quarrymen as ring stones.[14] The structures, from the faces of the North Quarry, Woodhouse Eaves in Charnwood Forest, Leicestershire, England, are now known to be genuine Ediacaran fossils. Ironically, it was in Africa, where the usually common medusoids are rare, that Ediacaran fossils were first described and interpreted as fossils.

The oldest medusoid, approximately 600 million years old,[15] was found by my field party in Sonora, Mexico (figure 2.1). The expedition leading to its discovery is the subject of chapter 9. Like the discovery of fossils in the Ediacara Hills of Australia, the Mexican find was an outgrowth of a government-sponsored effort (by the U.S. Geological Survey and the Mexican Recursos Minerales; color plate 1) to characterize the mineral resources of a remote area.

FIGURE 2.1: The oldest Ediacaran, a specimen of *Cyclomedusa* from the Clemente Formation of Sonora, Mexico. Fossil is viewed from the bottom. Greatest dimension of rock specimen is 6 cm.

The youngest Ediacaran fossils known are also medusoids (figure 2.2). Their discovery in Booley Bay, County Wexford, Ireland, led T. Peter Crimes to conclude that "there was no mass extinction at the end of the Precambrian."[16] The specimens were transported and deposited in a deepwater setting, although it is possible that they originated in shallow water and were carried downslope by submarine currents.

The oldest and youngest medusoid fossils are actually quite similar. Both have concentric and radial elements. Of most interest here is the tubular or flamelike nature of the radial elements, which are concentrated on the periphery of the organism.

A. Seilacher was first to point out that, unlike a modern jellyfish, the radial elements in an Ediacaran medusoid are on the outer edge of the specimen. In true jellyfish, the structures on the outer periphery of the body are concentric muscle bands that contract the jellyfish's bell and, when contractions are rhythmically synchronized, allow it to swim.

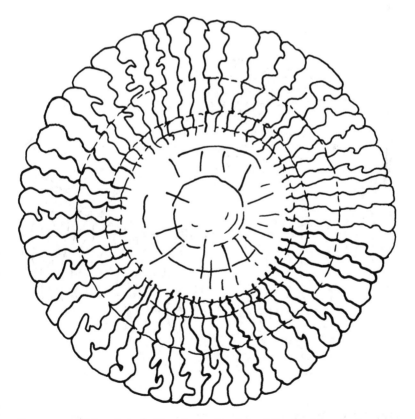

FIGURE 2.2: *Ediacaria booleyi* from Booley Bay, Ireland. Fossil is 13 cm in diameter.

Preston Cloud was one of the first to contest Seilacher's interpretation in print, though indirectly. In his last book Cloud illustrated the "medusoid phase of a modern aequorian hydroid" next to a photograph of a specimen of *Cyclomedusa.*[17] The strongest similarity between the two (and it is a striking similarity) is the resemblance between the radial features running from near the center of the underside of the modern medusa to its outer margin (figure 2.3) and the radial features in the *Cyclomedusa.*

There is a serious problem with Cloud's riposte to Seilacher, however. It is highly unlikely that the Ediacaran medusoids ever swam. They were benthic organisms that lived nested in the sediment. Early in the published records of the Australian finds, it was noted that the medusoids were preserved convex side down, like a bowl in the sediment. The bowls had reliefs of up to 15 mm, although actual relief in life, before sediment

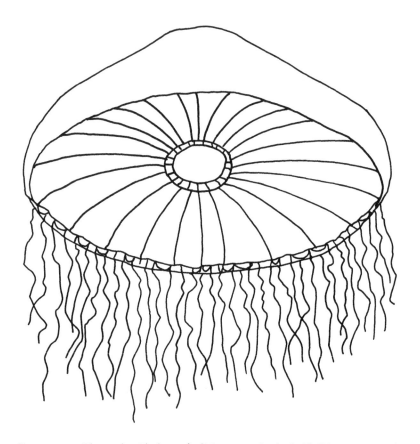

FIGURE 2.3: The medusoid phase of a living aequorian hydroid. Diameter approximately 1 cm.

compaction, would have been greater. In chapter 5 we see how recognition of the convex-side-down aspect of Ediacaran medusoids was the precursor to Garden of Ediacara theory, the concept that these medusoids were actually bowl-shaped solar collectors.

Many Ediacaran medusoids are three-dimensional fossils filled with fine sediment. The Mexican specimen (figure 2.1) is one of these, as is the Irish find. How did sand get inside these medusoids? Do they represent a sand casting of the impression in the sediment where a soft body once was, or was the sand part of the creature in life?

Uncertainties about this question have led to wide divergences in the interpretation of the most spectacular medusoid fossil, *Mawsonites* (figure 2.4). This ornate form, which has graced the covers of magazines in full color,[18] has been interpreted as a medusoid and as a trace fossil.[19] As noted in Stuart A. Baldwin's catalog of fossil reproductions, in the description of the reproduction of the holotype of *Mawsonites spriggi,*

> A superb specimen in very strong positive relief showing the many arcs of large irregular bosses which increase in size outwards towards the lobate periphery. . . . When originally described it was thought

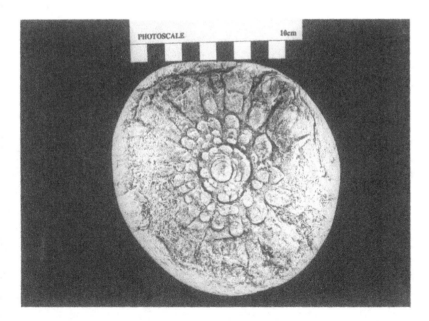

FIGURE 2.4: The Ediacaran medusoid *Mawsonites spriggi,* from the Ediacara Member of the Rawnsley Quartzite, Ediacara Hills, Flinders Ranges, South Australia. Scale bar in centimeters.

to be an unusual form of Medusa but Prof. A. Seilacher (personal communication, 1987) on seeing one of our replicas for the first time diagnosed it as a TRACE FOSSIL with a central burrow surrounded by backfill structures (the bosses).[20]

Seilacher published this suggestion in a scientific paper as well[21] but later abandoned the idea. Runnegar suggests that *Mawsonites* is a holdfast.[22]

I think that the problem with the interpretations discussed here is that, with the medusoids, we are dealing with sand creatures. I am not talking here about Seilacher's *Psammocorallia* concept.[23] Seilacher has inferred that certain discoid fossils formed of sand were the internal, organically cemented sand skeletons of a sea-anemone-like creature. He felt that this weighting of sand in the bottom of the organism would help the creature remain upright and act as an anchor that would cause the base of the anemone to be automatically implanted into sandy surfaces when it was rocked by currents (figure 2.5). I suggested to Seilacher the phrase "rock in a sock" to describe this arrangement, and he uses the phrase in his drawings of the psammocorals.[24]

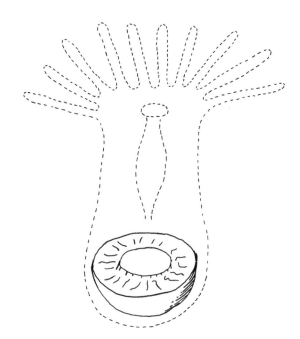

FIGURE 2.5: Seilacher's concept of the psammocoral, a sea-anemone-like creature with an internal sand skeleton. Diameter of sand skeleton 2 cm.

Ediacarans contained sand that pervaded their bodies, similar to the sand-filled character of the psammocorals. With the Ediacaran forms, it is as if the internal sand has been intimately incorporated through the body walls themselves.[25] The youngest known Ediacarans (figure 2.2) are apparent examples of this tendency. In 1995, T. Peter Crimes at the University of Liverpool and his colleagues described the Ediacaran fossil *Ediacaria booleyi* from Upper Cambrian rocks of the Booley Bay Formation in County Wexford, Ireland.[26] I doubt that these Cambrian forms actually belong in the medusoid genus *Ediacaria*,[27] but they do appear to be Ediacarans. The radial, fingerlike projections between concentric elements are very much like those seen on older Ediacarans, including the oldest known from Mexico. Crimes et al. interpret the Irish specimens as sand casts, but I believe that at least some of the sand now forming the domal fossils was once part of *E. booleyi*'s body.

This was an inverted domal or filled-bowl creature in life, and sand was very much a part of its body. This explains why as careful an observer as Seilacher could mistake *Mawsonites* for a trace fossil; like a trace fossil, this Ediacaran actually constitutes a significant chunk of the sediment itself.

Eoporpita (figure 2.6) has a body form that is emblematic of Ediacaran medusoids, a form that has been called the "tentaculate disc." It is essentially a discoid structure with radial elements ("tentacles") on its outer edges. Like the term *medusoid*, however, the term *tentaculate disc* is unfortunate because it assumes the unproved proposition that the radial structures were indeed tentacles.

An interesting observation can be made about the so-called tentaculate discs. The concentric markings look as if they might represent increments of growth. The medusoid *Mawsonites randellensis* of Bunyeroo Gorge, western Flinders Ranges, Australia (a large, 20-cm-diameter fossil), shows what might be concentric growth banding.[28] This is seen even more clearly in the Irish specimens of Booley Bay. In some specimens of *Ediacaria booleyi* the banding looks so orderly that it might be possible to pick out monthly or annual cycles.[29] In one specimen there is a clear alternation between "tentaculate" and concentric growth modes.

In an interesting paper published in 1993, Muscovite paleontologist Andrei Yu. Zhuravlev argued, following a thought first expressed by Hans D. Pflug, that Ediacaran fossils were not multicellular, like animals, but were giant unicells.[30] Zhuravlev argued, following Seilacher's

FIGURE 2.6: The medusoid Ediacaran *Eoporpita*. This type of Ediacaran is also known as a tentaculate disk. Specimen is 7 cm in diameter.

inference that Vendobionts may have been syncytial,[31] that Ediacaran forms were related to a modern group of giant protists[32] called the xenophyophores.

Xenophyophores are unfamiliar creatures living today in the deep sea. These giant deepwater marine unicells form a skeleton of agglutinated sediment particles. Their feeding strategy probably involves both digestion of organic matter in sediment (hence their preference for food-rich conditions on the seafloor) and direct absorption of nutrients from seawater. Nevertheless, Zhuravlev took his own theory with a grain of salt, saying that "full identification of Vendobionta with xenophyophores would seem to me a stretch."[33]

A new piece of information may support the xenophyophore model for at least some of the Ediacaran taxa. The same year as the publication of Zhuravlev's inferences about the antiquity of xenophyophores, Andrew J. Gooday of the Southampton Oceanography Centre in England reported, in the journal *Deep Sea Research I*, direct observations of episodic growth in an abyssal xenophyophore.[34] In this study, three specimens of the species *Reticulammina labyrinthica* were photographed over an 8-month observation period. During this interval, growth occurred in distinct, episodic intervals, each episode being separated from the previous by about 2 months.

Could these growth pulses in modern xenophyophores be an expression of a similar mode of growth in Ediacaran discoid organisms? In other words, were the concentric additions to the sediment-filled body of *Ediacaria booleyi* successive pulses of growth and sediment incorporation? If so, this would support Zhuravlev's hypothesis of a phylogenetic link between Ediacaran creatures and xenophyophores.

Figure 2.7 is a sketch made from a photograph published in 1991 by Jim Gehling in the *Geological Society of India Memoir*.[35] The sketch shows two clusters of individuals of the species *Cyclomedusa davidii*. The concentric structure of each individual is disturbed where it contacts the next. There is also a pattern in the off-center concentric bands. Compare the banding in the topmost right specimen (largest) and the leftmost specimen in the lower cluster. Both of these specimens have five incremental growth bands. I suggest that these growth bands represent synchronous increments of growth. Such a growth pattern is similar to that of a xenophyophore, but the fossil pattern is geometrically simple. Any number of organisms could generate growth patterns such as these.

FIGURE 2.7: The medusoid Ediacaran *Cyclomedusa davidii*, showing paired and clustered individuals. Diameter of the largest is 2.85 cm.

Sketch from plate 5, figure 1 of J. G. Gehling, "The Case for Ediacaran Fossil Roots to the Metazoan Tree," *Geological Society of India Memoir* 20 (1991):181–224.

Dumplings

The Ediacaran biota of the Mistaken Point Formation, Avalon Peninsula, Newfoundland, at the extreme easternmost point of North America, is one of the largest, best-exposed, and most accessible Ediacaran localities.[36] (Mistaken Point is named for fatal errors made by sailors who mistook it for nearby Cape Race [guidepost to safe harbor at St. John's] and ended up shipwrecked on the rocky coast.)[37]

Among the numerous fossils of Mistaken Point is (in addition to numerous frondose fossils) an enigmatic form called lobate discoidal remains (figure 2.8).[38] These have been nicknamed dumplings.

The modern xenophyophore *Reticulammina labyrinthica* is composed of agglutinated sediment particles very much like its surrounding sediment. If it were to preserve as a fossil, however, chances are that the fossil would look very much like the ancient dumplings, the lobate discoid remains of Newfoundland. These fossil specimens may therefore represent the remains of shallow-water xenophyophores.

Tribrachidium heraldicum

Tribrachidium (see figure 1.1) was introduced as the mystery fossil in chapter 1. As a radially symmetric form, it is fairly typical of members

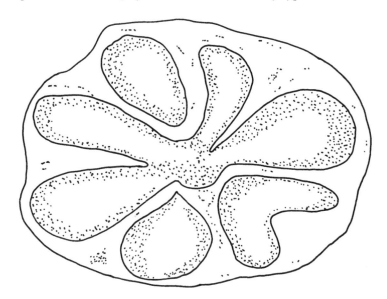

FIGURE 2.8: Lobate discoidal remains from the Ediacaran locality of the Mistaken Point Formation, Conception Group, eastern Newfoundland. Diameter 10 cm.

of the Ediacaran biota, and some have lumped it together with the medusoids. However, paleontologists have always been uneasy about the triradial symmetry of *Tribrachidium.*

Some of the medusoids, such as *Conomedusites,* have a distinctive fourfold or quadripartite symmetry, and this poses no difficulty for classifying them together with jellyfishlike animals. Modern jellyfish have this same type of symmetry, but very few animals have the true triradial symmetry seen in *Tribrachidium.*

Paleontologist Mikhail A. Fedonkin, who has made a career of studying the Ediacarans of the White Sea region of Russia, has focused attention on this threefold symmetry of *Tribrachidium.* Fedonkin compares it to triradial symmetry seen in certain types of tubular Early Cambrian shelly fossils (figure 2.9).[39] The similarities are interesting, but whether they point to anything more than a superficially shared triradial character is not known.

A close look at *Tribrachidium* and related genera such as *Albumares* and *Anfesta* (figure 2.10) shows them to be partitioned into three wedge-shaped to shoe-shaped sections. The former (wedge-shaped) can be transformed into the latter (shoe-shaped) by giving the central axis a clockwise twist. Thus these forms must be closely related by virtue of cognate morphology. I thus refer to *Tribrachidium, Anfesta,* and *Albumares* as tribrachidiids.

Examining a single triangular sector more closely shows it to be covered by fine striae that radiate from the apex of the triangle or wedge out to its outer edge. The striae bifurcate four or five times, so the outer edge appears to be marked by a fringed border.

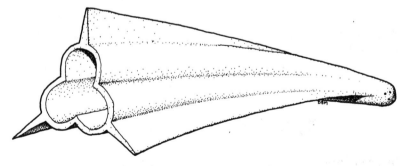

FIGURE 2.9: *Anabarites,* an Early Cambrian shelly fossil with triradial symmetry. The function of the vanelike stringers present in this species is unknown but may have acted to keep the shell from rolling with current. Length of shell 5 mm.

From M. A. S. and D. L. S. McMenamin, *The Emergence of Animals: The Cambrian Breakthrough* (New York: Columbia University Press, 1990). Artwork by Dianna McMenamin.

FIGURE 2.10: Two tribrachidiid Ediacarans. Left: *Albumares brunsae,* Ust-Pinega Formation, Summer Shore of the White Sea, Russia. 10 mm diameter. Right: *Anfesta stankovskii,* Ust-Pinega Formation, Winter Coast of the White Sea, Russia. 18 mm diameter.

The bifurcation of the striae, especially in *Albumares,* gives the lobes of the organism a distinctively fractal look. But the tribrachidiids are not the only Ediacaran forms with this pattern. Examine the Ediacaran form described here (figure 2.11; appendix) as *Gehlingia dibrachida.* This form has much in common with two of the lobes of a tribrachidiid, even down to the level of the side bulges along the axis. Such bulges are seen (figure 1.1) on each of *Tribrachidium's* three arms.[40] The bifurcating striae are clear also in *Gehlingia.*

Timing in the rate at which the striae bifurcate seems to be a characteristic of the tribrachidiids. In indifferently preserved specimens of *Tribrachidium,* all that is visible on the periphery of the specimen are the partitions between each of the three sets of striae. Rapid bifurcation leads to *Albumares.* Delayed bifurcation, forming tubular partitions between the striae, occurs in both *Tribrachidium* and *Gehlingia.* This is very much in accord with Seilacher's arguments that the Ediacaran body fossils are all related. I agree on this point with Seilacher, for the tribrachidiids and the related gehlingiids share essential morphological elements.

Phyllozoon *and the Frond Fossils*

In 1958 Trevor D. Ford presented a scientific paper about an unusual fossil organism from ancient sediments of Charnwood Forest, English Midlands.[41] The fossil was found in an abandoned quarry by a schoolboy named Roger Mason. Ford named the new form after its discoverer,

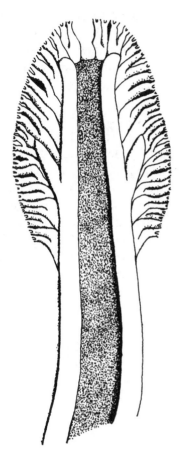

FIGURE 2.11: *Gehlingia dibrachida*, a new genus and species of Ediacaran. From the Ediacara Member of the Rawnsley Quartzite, Ediacara Hills, central Flinders Ranges, South Australia. Length of specimen 8.1 cm.

From M. A. S. McMenamin and D. L. S. McMenamin, 1994, *Hypersea: Life on Land* (New York: Columbia University Press). Modified from art by Heather Winkelmann.

calling it *Charnia masoni* (color plate 2). That same year, Martin F. Glaessner wrote "New Fossils from the Base of the Cambrian in South Australia," an early account of the Ediacaran fossils of Australia.[42]

Although these articles described fossils similar in age and overall aspect, they differed in a fundamental respect. Ford thought that *Charnia* was a photosynthetic alga, whereas Glaessner thought that the frondose forms from the Ediacara Hills were sea pens.

This disagreement made its way into the popular literature of paleontology. The following excerpt is from *Fossils: A Guide to Prehistoric Life* (A Golden Nature Guide) published in 1962, still in print, and widely read by kids of my generation: "*Charnia*. . . . Upper Pre-Cambrian, is a disputed fossil known from England and Australia. It is regarded by some as an alga and by others as a sea pen one of the coelenterates. Length 4 to 8 in."[43]

The editors of Golden Nature Guides apparently agreed with Ford, for *Charnia* is placed in the chapter on "Fossil Plants." The entry is now out-of-date, for the Australian fossil assigned in the excerpt earlier to *Charnia* is now placed in the genus *Charniodiscus*. Ford reluctantly abandoned his algal interpretation of *Charnia* in the face of Glaessner's vigorous promotion of hypothesis of animalian affinities.[44]

Another frond fossil (figure 2.12) is the form called *Phyllozoon*, first described as *Phyllozoon hanseni* in 1978 by Richard J. F. Jenkins and James G. Gehling. The species was named for Anthony Kym Hansen, who first discovered the fossils while studying geology at Adelaide

FIGURE 2.12: *Phyllozoon hanseni*, an Ediacaran frond from South Australia. Length of frond 13 cm.

University.[45] Hansen was killed during seismic exploration in Western Australia in 1976.

The genus name is derived from *phyllon*, the Greek noun for *leaf*, and *zoon*, Greek for *animal*, but it could alternatively be portrayed as a corrugate tongue. *Phyllozoon* is a typical example of a two-dimensional frond, with two vanes separated by the zigzag medial suture. Unlike the somewhat similar frond fossil *Pteridinium* (figure 2.13), *Phyllozoon* shows no evidence of a third vane emanating from the junction of the other two vanes.

Phyllozoon specimens may attain 25 cm or more in length, but they are by no means the largest Ediacaran fossils. Specimens of *Charniodiscus oppositus* attain sizes of well over a meter in length and at least 28 cm in width, which makes them the largest individual organisms known of their time. As far as we know, only microbial colonies called stromatolites (not all of which can be assumed to have been biogenic) reached larger size at this time.

Phyllozoon appears to have been a gregarious creature, for specimens are usually found in groups. One 1.2-m slab from the Ediacara Member of the Rawnsley Quartzite, central Flinders Ranges of South Australia, carries impressions of at least 11 individual specimens, closely associated with impressions of enigmatic large tube worms (figure 2.14) 3 cm in width and up to 50 cm in length. These tubular fossils have been compared with the giant vestimentiferan tube worms of modern deep-sea hydrothermal vents.[46]

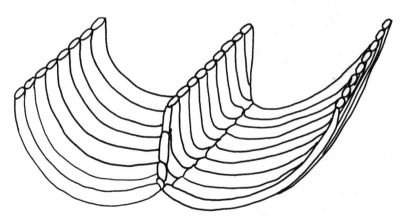

FIGURE 2.13: Diagrammatic sketch of a section of the frondose Ediacaran *Pteridinium*. The central vertical wall is informally called the chaperone wall, whereas the adjacent curving walls are called the bathtubs. Width of specimen 5 cm.

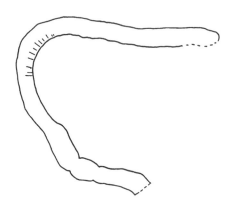

FIGURE 2.14: Large tubular fossil known to occur with Ediacarans in the Rawnsley Quartzite beds of South Australia. Possibly the remains of some type of tube worm. Diameter of tubes 15 mm.

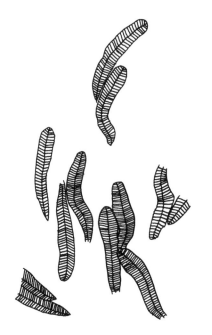

FIGURE 2.15: Sketch of lower surface of a bed of Rawnsley Quartzite (South Australia) showing numerous specimens of *Phyllozoon*. Note how many long axes of the specimens appear to be in approximate alignment.

Outline of fossils modified from B. Runnegar, "Proterozoic Eukaryotes: Evidence from Biology and Geology," in S. Bengtson, ed., *Early Life on Earth*, pp. 287–297 (New York: Columbia University Press, 1994).

Large *Phyllozoon* specimens can have up to 100 lateral grooves in each vane on either side of the zig-zag medial groove. This zigzag groove is an important morphological feature shared by many Ediacarans.

When separate fronds overlap, the grooves of one organism make a distinct overprint on the grooves of the adjacent frond. The shape of each leaf-shaped specimen includes a tapering stem end (figure 2.15). Specimens are often found preserved side by side, although the significance of this arrangement is not known. In some cases the juxtaposition of fronds is so close as to suggest that they might have been fused together, although this was probably not the case.

In the 1.2-m slab from the Flinders Ranges, the giant tube worms appear to be randomly oriented, but this is not the case for the *Phyllozoon* specimens. Figure 2.15 is an outline sketch of the Flinders slab, with positions of the *Phyllozoon* specimens shown. Note that the axes of many of the fronds are roughly parallel. Why would the axes be approximately aligned? Perhaps they were washed together and axis alignment proved to be a hydrodynamically stable orientation, in which case it would be accidental. But could there be another reason for the axis alignment?

Does such side-by-side alignment have something to do with *Phyllozoon*'s mode of reproduction? Was the next generation born aligned with the parent? Or did *Phyllozoon* have a preferred orientation, a positioning that would align itself with currents or maximize its ability to capture light?[47] The strap shape of the creature could simply be an artifact of its elongate growth pattern rather than having any particular adaptive value. But the orientation of the fronds, in what I presume was a sessile animal, must be telling us something about the creature's life habits.

What type of organism do these fossils represent? How did they grow? The reconstruction of the frond fossil *Thaumaptilon* from the Cambrian Burgess Shale is remarkably similar to models of *Charniodiscus* presented by the Australian school.[48] Does this indicate a true biological relationship, or was Conway Morris[49] unduly influenced by the conventional interpretations of Australian workers? Subsequent chapters focus on the debate surrounding these issues.

Erniettids

The erniettids are a group of Ediacarans best known from Namibia, although forms resembling erniettids are found in other parts of the world. The genus *Ernietta* was first described by Hans D. Pflug in 1966.[50] Pflug in 1972 followed this description with a published proliferation of ernio-genera to describe the various forms he saw represented by erniettid morphology.[51]

Pflug was roundly criticized for describing all these additional genera, with most paleontologists feeling that they were all synonymous with the original *Ernietta.*[52] In Pflug's defense, however, he was grappling with an exceptional organism that still defies all standard attempts at classification.

Ernietta is a bag-shaped creature composed of linked hollow tubes or parallel ribbing patterns that emanate from a zigzag medial suture, positioned like a zipper at the bottom of the bag. The narrow tubular chambers are in some cases filled with sand. Some specimens show a transverse constriction, or groove, perpendicular to the ribbing, that may be a zone of budding. Forms with the transverse constriction can be thick and squat, forming what has been called an elephant's foot. Erniettids in Namibia can be gregarious; one chunk of rock from the Kliphoek Member of the Dabis Formation, Nama Group, Namibia, shows a number of specimens closely juxtaposed in life position. Individual specimens are occasionally preserved with large pyrite crystals sticking out from the bottom of the bag.

Erniettids have been recently described from both Nevada[53] and Mexico.[54] Figure 2.16 shows the Nevada specimen. Each of the tubular chambers is filled by sand. It is hard to imagine how the sand could get into these tubes after the death of the organism, so I must interpret this specimen as evidence that erniettids had sand infillings in life.

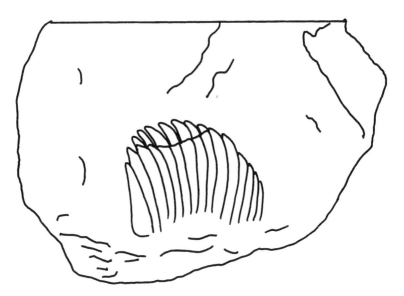

FIGURE 2.16: A specimen of an erniettid Ediacaran from southern Nevada. Note apparent damage and regrowth to top of specimen. Width of fossil 2.9 cm.

Rangea

The leaf-shaped fossil *Rangea schneiderhoehni* (figure 2.17) was first described by Gürich in 1929 from quartzites composed of coarse sand grains in the Kliphoek Member of the Dabis Formation. The type specimen was feared lost in World War II, when the museum in which it was held was struck by Allied bombers, but a salvage team picked the specimen from the smoking rubble. I examined this smoke-blackened specimen in 1993 at the home of Hans Pflug in Lich, Germany.

Rangea specimens have been recently found in Australia, indicating both that it was a widely distributed genus and that the Australian and Namibian biotas are similar in age. In 1982 I correctly predicted that *Rangea* would be found outside of Namibia.[55]

Richard Jenkins argues that *Rangea* is preserved mainly in storm-deposited sandstones. This does not accord with the contention of Seilacher that *Rangea* lived in the sediment. Seilacher and Jenkins also disagree on *Rangea's* overall body form. Jenkins sees the juxtaposition of multiple fronds, converging at the tip, as part of a single pleated frond.[56] Jenkins sees at least four fronds, up to six or more, combining to form

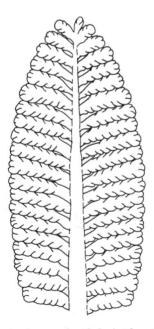

FIGURE 2.17: The Ediacaran *Rangea schneiderhoehni* from Namibia. Length of frond 11 cm.

the upper body of *Rangea*. The compound fronds of *Rangea*, and the tri-fold vanes of *Nasepia* and *Swartpuntia* from Namibia,[57] appear well suited to absorption of sunlight.

Dickinsonia

The best-known Ediacaran fossil, *Dickinsonia*, promises to provide a tremendous amount of information concerning the paleobiology of these organisms. If only we knew what it was. *Dickinsonia* has the distinction of being the only fossil to be described as a jellyfish, a coral, a sea anemone, an annelid worm,[58] a polychaete worm, an arthropod, a bac-terium, a protozoan, a member of a new phylum, a member of a new kingdom, and even an alien creature from outer space. In 1992 Rudolf Raff asked seven colleagues to identify what type of organism *Dickinsonia* was. He received seven different answers.[59]

My first encounter with specimens of *Dickinsonia* came as a graduate student in the early 1980s at the University of California at Santa Barbara in what was then the Biogeology Clean Lab.[60] The central cor-ridor of the building is lined with gray steel specimen cabinets, holding Cloud's collection of Cambrian and Precambrian fossils. Cloud had a few specimens of *Dickinsonia* from Australia. They weren't particularly good or complete specimens, but they did give a good sense of what such fossils were supposed to look like, permanently impressed into the yellowish-tawny colored Pound Quartzite.[61]

My next encounter came in 1988, when I convened, for the Ameri-can Association for the Advancement of Science, a symposium at the annual meeting in Boston titled "The Dawn of Animal Life or Aliens Here on Earth? Paleobiology of the Ediacaran Fauna." Richard Jenkins, whom I had invited to speak at the meeting, handed me a plaster cast of a small slab of quartzite bearing some of the best-known specimens of *Dickinsonia costata*. This cast is a marvelous thing to behold (figure 2.18); my students and I spend hours looking at it, trying to unleash its secrets. The slab has two large dickinsoniids, a number of juvenile spec-imens, and several specimens of the small Ediacaran *Parvancorina*.

Here is how a formal paleontological description of *Dickinsonia* might be written:

> *Dickinsonia* is oval in shape, broad and flat. It is bilaterally sym-metric, with a plane of symmetry bisecting the oval along its long axis. In some specimens there is a raised ridge running along the

FIGURE 2.18: *Dickinsonia costata* from the Flinders Range, South Australia. Length of specimen 13.4 cm.

symmetry line. Both halves of the creature are divided into tubular partitions that run approximately perpendicular to the plane of symmetry at the point where they meet the midline. Moving away from the midline, the tubular partitions become wider and curve gently toward the nearest end of the flat oval body. Adjacent tubular partitions are fused along their lateral edges for almost all of their length. At one end of the oval the tubes are long and constitute approximately one-third of the length of the midline of the organism. At the end opposite to this one, the tubular structures are much shorter. There may be an inward indentation in the perimeter of the oval at this end. The end of the oval with shorter tubules is presumably the end at which new tubular partitions are added during growth. The margins of the oval may show concentric wrinkling.

Four species of *Dickinsonia* are known: *D. costata, D. lissa, D. tenuis,* and now (with apologies to *Tyrannosaurus*) *D. rex.* Thus, dickinsoniids have more well-delineated species than any other member of the Australian Ediacaran assemblage.

D. rex is 43 cm long, appropriately described by Richard Jenkins as looking like a beaver's paddle. The type specimen of *D. rex* has a pronounced medial ridge, interpreted by Jenkins as the animal's lower intestinal tract, although in light of Seilacher's perspective this interpretation is controversial. Jenkins claims to be able to identify a fossil mouth in one dickinsoniid specimen, but the photograph has been retouched with pen and ink and it is difficult to say whether this single specimen is reliable.[62]

Much discussion has focused on the flexibility of *Dickinsonia* and its ability to exchange gas and nutrients through its cuticle. A major unsolved problem is how the cuticle could be thin and soft enough to be flexible and easily contracted, yet firm enough to stand up to the grains of a sandstone under the crushing pressures of rock lithification and, assuming it absorbed food, admissive of nutrients dissolved in water.

It is also difficult to determine which end of a *Dickinsonia* is its head and which is its tail, a major reason that the worm interpretation of its affinities has not been accepted by all paleontologists. Jenkins assumes that the end with the longest (and hence oldest) segments is the head or anterior end. However, one of the participants in my Chautauqua course ("The Ediacaran Biota" held May 18–20, 1995, Northern Illinois University, DeKalb), made the interesting suggestion that the enlarged tubes at the supposed anterior end of a dickinsoniid represent swollen gonads. Another suggestion is that these first two segments were modified to become sense organs. I discuss in chapter 11 why *Dickinsonia*'s reproduction probably occurred at the growing ("posterior") end.

Spriggina *and Soft-Bodied Trilobite*

Two types of Ediacaran fossils have what at first glance appears to be a "head" end. These are the genus *Spriggina* (figure 2.19) and an unnamed form called informally (and perhaps unfortunately) a soft-bodied trilobite. The putative head of *Spriggina* is a horseshoe-shaped termination of its supposed anterior end. No eyes are present on what has been called a cephalon, and a clever attempt some years ago to identify eyes on a digitized composite of specimens using image enhancement did not meet with success.[63]

Seilacher, as noted earlier, has criticized the annelid or arthropod interpretation of *Spriggina* and feels that it is simply another Vendobiont frond fossil with unipolar growth. Rudolf A. Raff has published an amusing cartoon of *Spriggina*, with the arthropod interpretation of this Ediacaran crawling past the frond interpretation of its form.[64]

Spriggina had what Seilacher would call unipolar growth. Forms with unipolar growth grow at only one end of the body; forms with bipolar growth grow at both ends. In this view, the anterior "cephalon" becomes

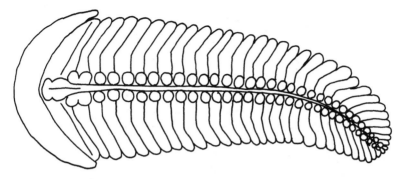

FIGURE 2.19: *Spriggina,* an Ediacaran from South Australia and Russia. Note the crescent moon–shaped structure at the widest end of the creature. Maximum width of organism 1.5 cm.

merely an enlarged, early subdivision of the body, perhaps doing service as a holdfast.

A form that appears to be tailor-made to perpetuate the Ediacaran controversy was announced by Jenkins and Gehling in the early 1990s. This so-called soft-bodied trilobite does indeed look strangely like a trilobite. The specimens have, on the "cephalon," what Jenkins interprets as "conspicuous recurved eye ridges" and a median glabella.[65] The rest of the body of the creature consists of a broad thorax of 21 segments, terminated by an "ovate pygidium . . . too small to show any detail in the single specimen in which it is preserved."

This creature (figure 2.20), represented by about seven specimens, has other features that are unusual for trilobites. Each individual segment of the body[66] expands distally (away from the midline), an exceedingly odd proportion for a trilobite. Also, each "pleural segment" has faint, regularly spaced lines perpendicular to the width (greatest dimension) of each segment. This feature is also unusual for trilobites, but is typical for Ediacarans such as *Charnia,* in which the striations are manifest as the secondary divisions of the quilts. Under Vendobiont theory, "soft-bodied trilobite" would probably be interpreted as a shortened or perhaps juvenile frond fossil. Indeed, there is no evidence of any limbs in the creature, so it really resembles a trilobite's shell or carapace, useless in itself for animal-style locomotion, and in any case not much like a true soft-bodied trilobite.[67] However, the soft-bodied trilobite cannot be a discarded shell because there is no evidence (splitting or breakage) that the body covering was abandoned by its maker.

Nevertheless, the fossil retains an eerily trilobite-like overall aspect, like a trilobite looking at itself in a funhouse mirror. If one could imagine a

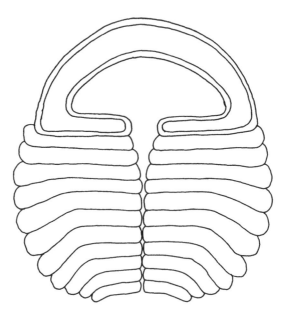

FIGURE 2.20: "Soft trilobite," an Ediacaran from South Australia. Note D-shaped ridge within the "head" region. 9 mm in length.

trilobite that was trying to expand its surface area to, say, absorb more light, it might indeed look something like this. What could this mean? Assuming that the form is a Vendobiont, with no close relationship to animals, what could the "head" end represent? Could it indeed be a head of sorts, with a concentration of sense organs with modes of operation and functions unfamiliar to us? Were the Vendobionts evolving heads and brains independently of animals? Iterative evolution of the brain? Or is it simpler to stick with an animalian interpretation for "soft-bodied trilobite," as well as for *Parvancorina*, a form that does indeed look like a larval trilobite, as suggested by Julian Kane and Joseph Cioni[68] (figure 2.21)?

In support of this idea, Richard Fortey et al.[69] compared *Parvancorina* with the larva of an unusual Lower and Middle Cambrian trilobite.[70] *Naraoia* resembles an overgrown trilobite larva and represents a developmental departure from normal trilobite growth known as hypermorphosis. Fortey et al. argue that the giant "protaspis" larva of *Naraoia* is so similar to *Parvancorina* that the latter may be an Ediacaran trilobite. It is as if Fortey and coauthors meant to say that the old evolutionary adage "ontogeny recapitulates phylogeny" is operating in reverse, with the hypermorphosis of *Naraoia* atavistically evoking a "parvancorinid" stage of development early in *Naraoia*'s own ontogeny.[71]

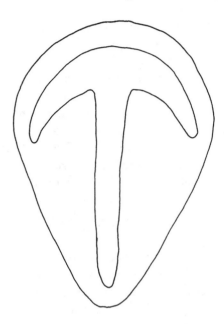

FIGURE 2.21: *Parvancorina,* a shield-shaped form from South Australia. 9.3 mm in length.

Jenkins has characterized the new theories (Vendobionta, Garden of Ediacara) as "surrealistic interpretations."[72] The fault is not of theorists, however, but of the surrealistic fossils themselves, which have proved impossible to interpret in a straightforward and wholly satisfactory fashion. What a fascinating puzzle!

Ediacaran Sponges

Gehling and Rigby argue in a recently published paper that their newly described body fossil, named *Palaeophragmodictya reticulata,* is the oldest known sponge.[73] For simplicity, I call this species *frag.*

Gehling and Rigby based this assignment on networklike impressions (figure 2.22) on the surface of the disc-shaped fossils and interpreted the network as impressions of linked, cross-shaped sponge spicules (called hexactinellid spicules). Perhaps because no trace (other than the supposed impressions) of the putative spicules remained in the fossil, Gehling and Rigby were remarkably candid concerning their own reservations about assigning these fossils to the sponges. They admitted that "this phylogenetic placement may be regarded as controversial."[74]

Again, this is a problem with interpretation, leaving investigators uncertain about even their own best judgment.

An interesting aspect of frag is that when (presumed) cohorts grew together, there seems to have been a preferred minimum spacing between individuals (figure 2.23). The minimum spacing appears to be approximately 3 mm (with one exception). Perhaps the evenness of the spacing is caused by the outer flange of each individual pushing its nearest neighbors away with continued growth. This would be a useful adaptation to maximize light capture and avoid shading, assuming (as seems likely) that these domal forms were photosymbiotic.

Ausia

This curious form, *Ausia fenestrata*, is named for the town of Aus, Namibia, and for the fact that it appears to have windows (Latin *fenestra*, "window"). *Ausia* is shaped like a hollow cylinder with a conical end that tapers to a point. The surface of the cylinder (figure 2.24) is marked by what appear to be rows of large pores. The pores become elongate and narrow near the tip of the cone.

This may be the only Ediacaran known to have pores,[75] although Seilacher (personal communication, 1993) has argued that the "windows" are merely dimples and do not go all the way through the cuticle. Another possibility is that the windows are enclosed pockets for storage of food or symbionts.

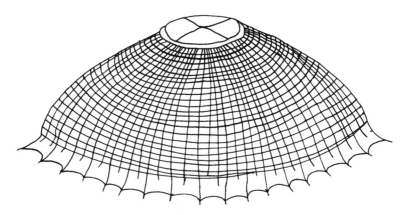

FIGURE 2.22: *Palaeophragmodictya reticulata* (alias *frag*), a dome-shaped organism from South Australia. Diameter 6 cm.

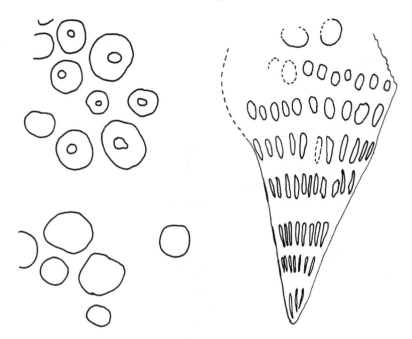

FIGURE 2.23: *Palaeophragmodictya reticulata* spacing on a rock slab from the Chase Range, South Australia. Greatest dimension of individual organism shown is 2 cm.

Sketched from figure 6-1 of J. G. Gehling and J. K. Rigby, "Long Expected Sponges from the Neoproterozoic Ediacara Fauna of South Australia," *Journal of Paleontology* 70 (1996):185–195.

FIGURE 2.24: *Ausia fenestrata,* an odd Ediacaran from the Nama Group of Namibia. Length of organism 5 cm.

If the pores are indeed holes, then *Ausia* might be linked to the sponges, although there is no evidence of spicules in this fossil.

Corumbella

Corumbella holds the record for the toughest integument of an Ediacaran (except the mineralized cloudinids, to be discussed in chapter 6). Described in 1982, *Corumbella werneri* (figure 2.25) forms deep impressions, showing a central ridge and perpendicular ridgelets.[76] It may be a frond fossil with a particularly heavy integument.

Its description inspired me to suggest *Corumbella* as a link between the Ediacaran frond fossils and an enigmatic group of Paleozoic organisms known as conularids.[77] This idea merits serious consideration. Conularids, like Ediacaran fronds and *Corumbella,* have a flexible integument and one or more zigzag medial sutures. The large conularid *Paraconularia chesterensis* bears a pronounced morphological similarity to *Pteridinium*, an Ediacaran frond that will be described in greater detail in chapter 4.

Inaria karli

Inaria karli (figure 2.26), from the classic Ediacara localities in Australia, resembles a fig or an intact cluster of garlic cloves. The lower part of the organism is divided into radially arranged partitions, which converge upward to form a tubular funnel. Fossils of the form are now all flattened, but *Inaria karli* was originally three-dimensional. Gehling,

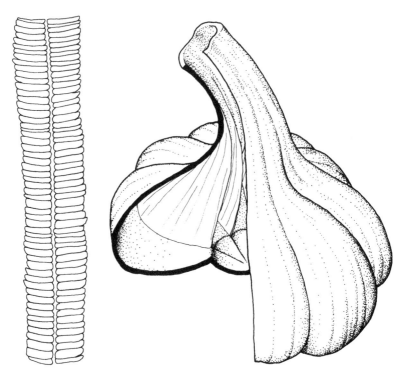

FIGURE 2.25: *Corumbella werneri*, a fossil from the Corumbá Group of Mato Grosso, Brazil. Specimen is 4.5 mm in width.

FIGURE 2.26: *Inaria karli* has a sac-shaped body plan. This figure shows a cutaway of a mature specimen in life position. Greatest width of specimen about 7.5 cm.

From M. A. S. and D. L. S. McMenamin, *The Emergence of Animals: The Cambrian Breakthrough* (New York: Columbia University Press, 1990). Artwork by Dianna McMenamin.

who described the species, suggests that *Inaria karli* had a bag-shaped body plan designed for photosymbiosis, and may have acted as a "respiring culture-chamber."[78] Strategies such as these may have rendered *Inaria* independent of external, digestible, living sources of food.

Arkarua adami

Arkarua adami (figure 2.27), also described by Jim Gehling,[79] is a globular fossil with five rays on its surface and an outer flange with radial markings. Based on its putative similarities to an ancient type of discoid echinoderm,[80] *Arkarua adami* is presented by Gehling as the oldest known fossil echinoderm. *Arkarua's* similarities to echinoderms of later times may prove superficial, however. Consider the similarities between the radially marked flange in *Arkarua adami* and (with apologies for the alliteration) the fringing flange in frag. A fourfold radial marking is seen in the middle of porelike spaces seen on the top of frag. *Arkarua* could be a variant of frag.

Gehling consistently favors the shoehorn approach into modern *animal* taxa for his newly described Ediacaran forms. Doing so is more acceptable to conservative paleontologists, who seem to feel that it somehow strengthens the reputation of invertebrate paleontology. In some cases this approach may bring the right answers, but as Seilacher has so aptly pointed out, other approaches must also be tried if we are to arrive at a full understanding of the Ediacarans.

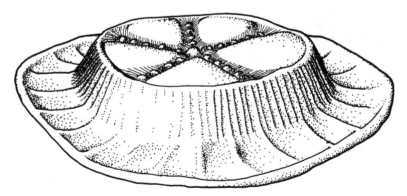

Figure 2.27: *Arkarua adami*, an Ediacaran with fivefold (pentameral) symmetry from South Australia. Diameter of fossil 6 mm.

From M. A. S. and D. L. S. McMenamin, *The Emergence of Animals: The Cambrian Breakthrough* (New York: Columbia University Press, 1990). Artwork by Dianna McMenamin.

Notes

1. D. C. Dennett, *Consciousness Explained* (Boston: Little, Brown, 1991).
2. See p. 389 in E. Emmons, "On New Fossil Corals from North Carolina," *American Journal of Science (new series)* 22 (1856):389–390.
3. M. A. S. McMenamin and D. L. S. McMenamin, *The Emergence of Animals: The Cambrian Breakthrough* (New York: Columbia University Press, 1990).
4. See figure 1b in E. Emmons, "On New Fossil Corals from North Carolina," *American Journal of Science (new series)* 22 (1856):389–390.
5. Pseudofossils are the scourge of Proterozoic paleontology. Compare the ancient Ediacaran "ring medusoid" *Bunyerichnus dalgarnoi* (figure 2F in P. Cloud and M. F. Glaessner, "The Ediacaran Period and System: Metazoa Inherit the Earth," *Science* 218 [1982]:783–792) with the structure formed by a tethered piece of wind-blown grass (B. Snyder, "First Prize—Plants and Their Environment," *Natural History* 91 [1982]:28).
6. Gold was sufficiently abundant that a mint for U.S. gold coinage was operated in nearby in Charlotte, North Carolina.
7. P. Cloud, "Pseudofossils: A Plea for Caution," *Geology* 1 (1973):123–127.
8. A nongeneric term such as *disc* would have been better.
9. R. C. Sprigg, "Early Cambrian (?) Jellyfishes from the Flinders Ranges, South Australia," *Transactions of the Royal Society of South Australia* 71 (1947):212–224; R. C. Sprigg, "Early Cambrian 'Jellyfishes' of Ediacara, South Australia, and Mount John, Kimberley District, Western Australia," *Transactions of the Royal Society of South Australia* 73 (1949):72–99.

Sprigg was a perceptive geologist. In a paper published in the early 1950s (R. C. Sprigg, "Sedimentation in the Adelaide Geosyncline and the Formation of the Continental Terrace," in M. F. Glaessner and E. A. Rudd, eds. *Sir Douglass Mawson Anniversary Volume, Contributions to Geology*, pp. 153–159 [Adelaide, Australia: University of Adelaide, 1952]), Sprigg correctly identified the Ediacaran fossil-bearing South Australian sequence of strata as a (p. 159) "fossil continental terrace" or continental shelf, and thus became the first geologist to recognize such a thing (P. F. Hoffman, personal communication). Sprigg also anticipated plate tectonic theory, noting that "geosynclinal downwarping" must be secondary to some other "*over-ruling tendency*" (p. 159, italics mine).
10. M. F. Glaessner, "New Fossils from the Base of the Cambrian in South Australia," *Transactions of the Royal Society of South Australia* 81 (1958):185–188.
11. H.-H. Vogt, "Fossilienschmuggel in Australien," *Naturwissenschaftliche Rundschau* 46 (1993):119.
12. See p. 757 in E. Hill and T. G. Bonney, "The Precarboniferous Rocks of Charnwood Forest," *Quarterly Journal of the Geological Society of London* 33 (1877):754–789.
13. R. J. F. Jenkins, "Functional and Ecological Aspects of Ediacaran Assemblages," in J. H. Lipps and P. W. Signor, eds., *Origin and Early Evolution of the Metazoa*, pp. 131–176 (New York: Plenum, 1992).
14. B. Runnegar, "Vendobionta or Metazoa? Developments in Understanding

the Ediacara 'Fauna,' " *Neues Jahrbuch für Geologie und Paläontologie, Abhandlungen* 195, nos. 1–3 (1995):303–318.

15. The age was determined using the trace fossil *Vermiforma*, discussed in chapter 3; see M. A. S. McMenamin, "Ediacaran Biota from Sonora, Mexico," *Proceedings of the National Academy of Sciences USA* 93 (1996):4990–4993.

16. As cited in R. Monastersky, "Living Large on the Precambrian Planet," *Science News* 149 (1996):308.

17. See p. 316 of P. Cloud, *Oasis in Space* (New York: W. W. Norton, 1988).

18. P. Cloud and M. F. Glaessner, "The Ediacaran Period and System: Metazoa Inherit the Earth," *Science* 218 (1982):783–792.

19. A. Seilacher, "Late Precambrian and Early Cambrian Metazoa: Preservational or Real Extinctions?" in H. D. Holland and A. F. Trendall, eds., *Patterns of Change in Earth Evolution*, pp. 159–168 (Berlin: Springer-Verlag, 1984).

20. See p. 68 of S. A. Baldwin and P. F. Baldwin, *Educational Palaeontological Reproductions, Catalogue 5* (Essex, England: Stuart A. Baldwin, 1994).

21. A. Seilacher, "Late Precambrian and Early Cambrian Metazoa: Preservational or Real Extinctions?" in H. D. Holland and A. F. Trendall, eds., *Patterns of Change in Earth Evolution*, pp. 159–168 (Berlin: Springer-Verlag, 1984).

22. See p. 308 in B. Runnegar, "Vendobionta or Metazoa? Developments in Understanding the Ediacara 'Fauna,' " *Neues Jahrbuch für Geologie und Paläontologie, Abhandlungen* 195, nos. 1–3 (1995):303–318.

23. A. Seilacher and R. Goldring, "Class Psammocorallia (Coelenterata, Vendian Ordovician): Recognition, Systematics, and Distribution," *Geologiska Föreningens i Stockholm Förhandlingar* 118 (1996):207–216.

24. Jensen incorrectly interprets psammocorals as pseudofossils: A. Seilacher and R. Goldring, "Class Psammocorallia (Coelenterata, Vendian Ordovician): Recognition, Systematics, and Distribution," *Geologiska Föreningens i Stockholm Förhandlingar* 118 (1996):207–216. See also S. Jensen, "Trace Fossils, Body Fossils and Problematica from the Lower Cambrian Mickwitzia Sandstone, South Central Sweden," unpublished Ph.D. thesis, University of Uppsala, 1993.

25. As suggested by F. Pflüger.

26. T. P. Crimes, A. Insole, and B. P. J. Williams, "A Rigid-Bodied Ediacaran Biota from Upper Cambrian Strata in Co. Wexford, Eire," *Geological Journal* 30 (1995):89–109.

27. The Irish specimens differ considerably from the Australian material.

28. See p. 160 in R. J. F. Jenkins, "Functional and Ecological Aspects of Ediacaran Assemblages," in J. H. Lipps and P. W. Signor, *Origin and Early Evolution of the Metazoa*, pp. 131–176 (New York: Plenum, 1992).

29. See plate 6C in T. P. Crimes, A. Insole, and B. P. J. Williams, "A Rigid-Bodied Ediacaran Biota from Upper Cambrian Strata in Co. Wexford, Eire," *Geological Journal* 30 (1995):89–109.

30. A. Yu. Zhuravlev, "Were Ediacaran Vendobionta Multicellulars?" *Neues Jahrbuch für Geologie und Paläontologie, Abhandlungen* 190 (1993):299–314.

31. *Syncytial* is a term that refers to organisms with multiple nuclei in a shared (as opposed to multicellular) cytoplasm.

32. Protists are unicellular (or sometimes syncytial) organisms with nucleated cells (eukaryotes).

33. Zhuravlev, 1993, p. 309.

34. A. J. Gooday, B. J. Bett, and D. N. Pratt, "Direct Observation of Episodic Growth in an Abyssal Xenophyophore (Protista)," *Deep-Sea Research I* 40 (1993):2131–2143.

35. J. G. Gehling, "The Case for Ediacaran Fossil Roots to the Metazoan Tree," *Geological Society of India Memoir* 20 (1991):181–224.

36. W. K. Sacco, *Discovery* 23 (1992):36 shows a photograph of Gene Carrozza's marvelous Plexiglas sculpture of a Mistaken Point Ediacaran. This sculpture also graces the cover of A. Seilacher, *Fossile Kunst: Albumblätter der Erdgeschichte* (Korb, Germany: Goldschneck-Verlag, Werner K. Weidert, 1995).

37. K. Wright, "When Life Was Odd," *Discover* 18 (1997):52–61.

38. R. J. F. Jenkins, "Functional and Ecological Aspects of Ediacaran Assemblages," in J. H. Lipps and P. W. Signor, *Origin and Early Evolution of the Metazoa*, pp. 131–176 (New York: Plenum, 1992).

39. M. A. Fedonkin, "Precambrian Problematic Animals: Their Body Plan and Phylogeny," in A. Hoffman and M. H. Nitecki, eds., *Problematic Fossil Taxa*, pp. 59–67 (New York: Oxford University Press, 1986).

40. Or legs, if one compares it with the triskeles emblem of the Isle of Man.

41. T. D. Ford, "Pre-Cambrian Fossils from Charnwood Forest," *Proceedings of the Yorkshire Geological Society* 31 (1958):211–217.

42. M. F. Glaessner, "New Fossils from the Base of the Cambrian in South Australia," *Transactions of the Royal Society of South Australia* 81 (1958):185–188.

43. See p. 150 of F. H. T. Rhodes, H. S. Zim, and P. R. Shaffer, *Fossils: A Guide to Prehistoric Life (A Golden Nature Guide)* (New York: Golden Press, 1962).

44. B. Runnegar, "Vendobionta or Metazoa? Developments in Understanding the Ediacara 'Fauna,' " *Neues Jahrbuch für Geologie und Paläontologie, Abhandlungen* 195, nos. 1–3 (1995):303–318.

45. R. J. F. Jenkins and J. G. Gehling, "A Review of the Frond-like Fossils of the Ediacara Assemblage," *Records of the South Australian Museum* 17 (1978):347–359.

46. B. Runnegar, "Proterozoic Eukaryotes: Evidence from Biology and Geology," in S. Bengtson, ed., *Early Life on Earth*, pp. 287–297 (New York: Columbia University Press, 1994).

47. M. A. S. McMenamin and L. Bennett, "Light Transmission Through Sand: Implications for Photosynthetic Psammophiles and *Pteridinium*," *Geological Society of America Abstracts with Programs* 26 (1994):A-54.

48. See p. 313 in B. Runnegar, "Vendobionta or Metazoa? Developments in Understanding the Ediacara 'Fauna,' " *Neues Jahrbuch für Geologie und Paläontologie, Abhandlungen* 195, nos. 1–3 (1995):303–318.

49. S. Conway Morris, "Ediacaran-like Fossils in Cambrian Burgess Shale-Type Faunas of North America," *Palaeontology* 36 (1993):593–635. Conway Morris later admitted that the similarities between *Thaumaptilon* and Ediacarans could be a result of convergent evolution ("Why Molecular Biology Needs Paleontology," *Developmental Supplement* (1994):1–13.

50. H. D. Pflug, "Neue Fossilreste aus den Nama-Schichten in Südwest-Afrika," *Paläontologische Zeitschrift* 40 (1966):14–25.

51. H. D. Pflug, "Systematik der jung-präkambrischen Petalonamae Pflug 1970," *Paläontologische Zeitschrift* 46 (1972):56–67. Described by Pflug were *Erniofossa, Ernionorma, Erniobeta, Erniograndis*—a total of 12 new genera and 21 new species for what many other paleontologists believe are merely variants of *E. plateauensis* with different degrees of preservation.

52. Martin Glaessner scolded Pflug for this on p. A96 of M. F. Glaessner, "Biogeography and Biostratigraphy—Precambrian," in R. A. Robinson and C. Teichert, eds., *Treatise on Invertebrate Paleontology, Part A, Introduction, Fossilization (Taphonomy), Biogeography and Biostratigraphy*, pp. A79–A118 (Boulder, Colo., and Lawrence, Kans.: The Geological Society of America and the University of Kansas, 1979).

53. R. J. Horodyski, "Late Proterozoic Megafossils from Southern Nevada," *Geological Society of America Abstracts with Programs* 23 (1991):A163.

54. M. A. S. McMenamin, "Ediacaran Biota from Sonora, Mexico," *Proceedings of the National Academy of Sciences USA* 93 (1996):4990–4993.

55. M. A. S. McMenamin, "A Case for Two Late Proterozoic–Earliest Cambrian Faunal Province Loci," *Geology* 10 (1982):290–292.

56. R. J. F. Jenkins, "The Enigmatic Ediacaran (Late Precambrian) Genus *Rangea* and Related Forms," *Paleobiology* 11 (1985):336–355.

57. J. P. Grotzinger, S. A. Bowring, B. Z. Saylor, and A. J. Kaufman, "Biostratigraphic and Geochronologic Constraints on Early Animal Evolution," *Science* 270 (1995):598–604.

58. P. Cloud, "Pseudofossils: A Plea for Caution," *Geology* 1 (1973):123–127.

59. K. Wright, "When Life Was Odd," *Discover* 18 (1997):52–61.

60. Now the Preston Cloud Research Laboratory. This laboratory is at the University of California, Santa Barbara.

61. Now called the Rawnsley Quartzite.

62. See figure 12B in R. J. F. Jenkins, "Functional and Ecological Aspects of Ediacaran Assemblages," in J. H. Lipps and P. W. Signor, eds., *Origin and Early Evolution of the Metazoa*, pp. 131–176 (New York: Plenum, 1992).

63. J. L. Kirschvink, R. Kirk, and J. J. Sepkoski, Jr., "Digital Image Enhancement of Ediacaran Fossils: A First Try," *Geological Society of America Abstracts with Programs* 15 (1982):530.

64. See figure 3.3, p. 84 (drawn by E. C. Raff) in R. A. Raff, *The Shape of Life: Genes, Development, and the Evolution of Animal Form* (Chicago: University of Chicago Press, 1996).

65. See p. 169 of R. J. F. Jenkins, "Functional and Ecological Aspects of Ediacaran Assemblages," in J. H. Lipps and P. W. Signor, eds., *Origin and Early Evolution of the Metazoa*, pp. 131–176 (New York: Plenum Press, 1992).

66. Or "pleural segment."

67. For example, the Middle Cambrian *Marella*.

68. J. Kane and J. Cioni, "Precambrian Ancestors," *American Scientist* 75 (1987):569–570. Martin Glaessner also mentions a possible link between

Parvancorina and trilobites on p. A105 of M. F. Glaessner, "Biogeography and Biostratigraphy Precambrian," in R. A. Robinson and C. Teichert, eds., *Treatise on Invertebrate Paleontology, Part A, Introduction, Fossilization (Taphonomy), Biogeography and Biostratigraphy,* pp. A79–A118 (Boulder, Colo., and Lawrence, Kans.: The Geological Society of America and the University of Kansas, 1979).

69. R. A. Fortey, D. E. G. Briggs, and M. A. Wills, "The Cambrian Evolutionary 'Explosion': Decoupling Cladogenesis from Morphological Disparity," *Biological Journal of the Linnaean Society* 57 (1996):13–33.

70. H. B. Whittington, "The Middle Cambrian Trilobite *Naraoia,* Burgess Shale, British Columbia," *Philosophical Transactions of the Royal Society, London* B280 (1977):409–443.

71. B. Ekstig, "Condensation of Developmental Stages and Evolution," *Bio-Science* 44 (1994):158–164.

72. See p. 133 of R. J. F. Jenkins, "Functional and Ecological Aspects of Ediacaran Assemblages," in J. H. Lipps and P. W. Signor, eds., *Origin and Early Evolution of the Metazoa,* pp. 131–176 (New York: Plenum Press, 1992).

73. J. G. Gehling and J. K. Rigby, "Long Expected Sponges from the Neoproterozoic Ediacara Fauna of South Australia," *Journal of Paleontology* 70 (1996):185–195.

74. Gehling and Rigby, 1996, p. 185.

75. G. Hahn and H. D. Pflug, "Polypenartige Organismen aus dem Jung-Präkambrium (Nama-Gruppe) von Namibia," *Geologica et Palaeontologica* 19 (1985):1–13.

76. G. Hahn, R. Hahn, O. H. Leonardos, H. D. Pflug, and D. H. G. Walde, "Körperlich erhaltene Scyphozoen-Reste aus dem Jungpräkambrium Brasiliens," *Geologica et Palaeontologica* 16 (1982):1–18.

77. M. A. S. McMenamin, "The Fate of the Ediacaran Fauna, the Nature of Conulariids, and the Basal Paleozoic Predator Revolution," *Geological Society of America Abstracts with Programs* 19 (1987):29.

78. J. G. Gehling, "A Cnidarian of Actinian-Grade from the Ediacaran Pound Subgroup, South Australia," *Alcheringa* 12 (1988):299–314.

79. J. G. Gehling, "Earliest Known Echinoderm—a New Ediacaran Fossil from the Pound Subgroup of South Australia," *Alcheringa* 11 (1987):337–345.

80. Called an edrioasteroid; S. M. Stanley, "Fossil Data and the Precambrian-Cambrian Evolutionary Transition," *American Journal of Science* 276 (1976):56–76.

3 · *Vermiforma*

The worst problem in the search for the oldest animal fossils is mistaken identity.

—Mark and Dianna McMenamin[1]

On February 15, 1982, as field assistant to Jack Stewart of the U.S. Geological Survey, I picked up what appeared to be an unusual fossil. Our task for the day had been to make a reconnaissance hike through the Proterozoic sedimentary strata of Cerros de la Ciénega ("Hills of the Wellspring") in Sonora, Mexico.

Figure 3.1 shows a copy of my field notes for the day. We began the work by hiking down to the basement contact, which is the point in the sedimentary pile where the stratified rocks rest on the underlying granitic and metamorphic rocks. The basement rocks, in this case the igneous and metamorphic rocks of the Bamori Group, were long ago (tens of millions of years ago) exposed to air and water and erosion. The erosion was paused, however, when the first sediments of the Mexican Proterozoic sequence were deposited on the basement rocks, covering and protecting them like a blanket. The protective blanket remained until the entire sequence, crystalline basement rocks as well as overlying sediments, was tilted by tectonic forces and brought to the surface to weather away in the Sonoran desert.

We did not measure the exact thicknesses of the stratal layers encountered that day, but I did make ballpark estimates of each layer's thickness using my pace as an approximate yardstick. Here, in reverse stratigraphic order, are my descriptions of the rocks we encountered February 15th.

APPROXIMATE, EYEBALLED THICKNESSES

BAMORI: GRANITES, GNEISSES,[2] MIGMATITE,[3] GREENSTONE[4] NEAR UPPER CONTACT

5 METERS	Thin-bedded, cross-bedded reddish quartzite at contact
4 METERS	White cross-bedded sandstone
10 METERS	Interbedded thin to medium beds of tan silty dolostone and laminated purplish gray micro-cross-bedded quartzite

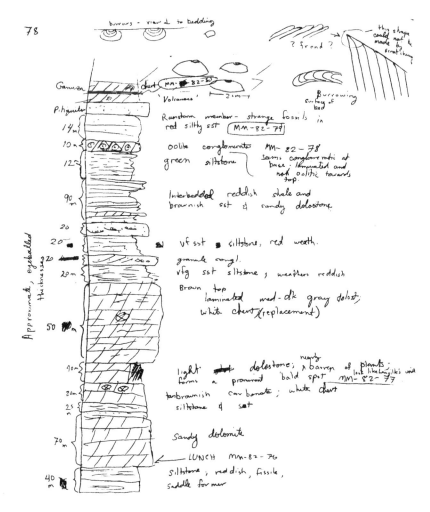

FIGURE 3.1: A page from my field notes of February 15, 1982, the day Jack Stewart and I measured a reconnaissance stratigraphic section in the Cerros de la Ciénega, Sonora, Mexico. Note the discovery of Ediacaran fossils at the top of the page.

20 METERS Dark gray dolostone, laminated, cross-bedded, thin-bedded; ridge former; silicified oolites?

5 METERS Shale

150 METERS Medium bluish gray laminated to thin-bedded dolostone; massive ridge former (near top of this unit) tan dolostone, silty; very fine-grained sandstone beds, granule conglomerates

10 METERS — Siltstone and quartzite

80 METERS — Light gray dolostone; indistinctly laminated; medium-bedded; scattered carbonate chips; white chert (top of this unit = top of El Arpa Formation?)

40 METERS — Siltstone, reddish, fissile, saddle former

LUNCH — MM-82–76

70 METERS — Sandy dolomite

25 METERS — Siltstone and sandstone

20 METERS — Tan brownish carbonate; white chert

40 METERS — Light-colored dolostone; nearly barren of plants; forms a prominent bald spot

50 METERS — Laminated medium to dark gray dolostone; white chert (replacement) brown top of unit

20 METERS — Very fine-grained sandstone and siltstone; weathers reddish

20 METERS — Granule conglomerate

20 METERS — Cross-bedded sandstone

90 METERS — Interbedded reddish shale and brownish sandstone and sandy dolostone

12 METERS — Green siltstone

10 METERS — Oolite conglomerates MM-82–78 seems conglomeratic at base; laminated and not oolitic toward top

14 METERS — Rainstorm member[*]—strange fossils in red silty sandstone

[*]Reference here is made to the Rainstorm Member of the Johnnie Formation, a correlative sedimentary unit in the United States.

MM-82–79 (here I made sketches). Volcanoes; burrowing surface of bed; burrows viewed perpendicular to bedding; frond?; this shape could not be made by scratching

No thickness
measured   Pitiquito (quartzite)

No thickness
measured   Gamuza (Formation); chert MM-82–80

When I picked up sample MM-82–79 and handed it to Jack Stewart, he said something like, "I've seen a lot of problematic sedimentary structures in this rock unit, but this one really looks like a trace fossil." Needless to say, I was quite pleased with my find for the day. However, the work of interpretation was just beginning.

The rock, a reddish-colored sandy siltstone, contained four elongate lobe-shaped objects, each with concentric U-shaped ridges at the end of the lobe (figures 3.2 and 3.3). The U-shaped ridges looked a lot like the concentric layers of sediment that form in many types of animal-built burrows.

When I returned to Santa Barbara I showed the objects to Preston Cloud. Cloud, who by 1982 had acquired a ferocious reputation as the premier debunker of Precambrian "fossils," agreed that they could be biologic. Encouraged by this, I published a photograph of these structures in the journal *Geology*, describing them as "probable metazoan traces."[5,6]

After publication I returned to Sonora and attempted to find more specimens of convincing trace fossils in the Clemente Formation, the rock unit that had produced these specimens, but without success. This was not an idle quest; I had my professors and fellow graduate and undergraduate students, both Mexican and American, engaged in searching and splitting Clemente Formation siltstones and sandstones. We simply struck out.

I showed the 1982 specimens again to Cloud. He still felt that they could be biologic, although he was intrigued by the fact that all four of the lobes were oriented in approximately the same direction. This made him suspicious, but he wasn't sure why.

In 1984, while I was completing my doctoral dissertation, Australian paleontologist Malcolm R. Walter visited Santa Barbara. I showed him the enigmatic structures, and he immediately came up with a way to form these lobes without invoking animal activity. Walter had seen flow

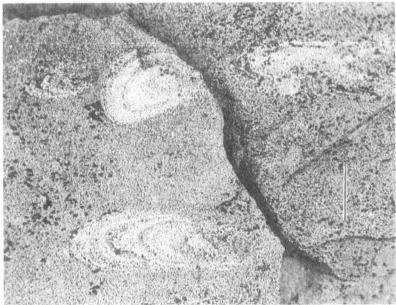

FIGURE 3.2: *Vermiforma* from the Clemente Formation of the Cerros de la Ciénega, Sonora, Mexico. Motion of the organisms (three individuals represented) was from left to right. Scale bar = 1 cm.

FIGURE 3.3: Enlargement of two of the specimens in figure 3.2. Scale bar = 1 cm.

structures resembling tiny lava flows coming off of the flanks of volcanoes, forming in association with sediment fluid-escape cones called sand volcanoes.

Sand volcanoes can form when water-saturated sediment is exposed to air and then disturbed by compactional forces or jostled by earthquakes. When this occurs, the sediment settles and forces water to move upward. The water sometimes follows a cylindrical conduit roughly resembling the vent of an igneous volcano. Sediment entrained in the water stream is deposited where the dewatering flow meets the air and can be deposited in a broad sediment cone or sand volcano. These sand volcanoes are often only a few centimeters in diameter, much smaller than their igneous counterparts. At the center of the sand mound is a small collapse pit, which looks like the vent, or caldera, in an igneous volcano.

When small sand volcanoes are preserved in ancient sediments, they are called pit-and-mound structures. Sometimes a particularly fluid slurry is ejected from the sand volcano vent. This slurry can flow down and beyond the flanks of the sand volcano, forming a lobe of fluidized sediment that can settle to create a sedimentary structure. This structure looks very much like a trace fossil with characteristic backfilled layers.

This was the interpretation I settled on for my dissertation text. Dianna McMenamin and I also opted for it in our 1990 book *The Emergence of Animals: The Cambrian Breakthrough*. Bruce Runnegar intimated that he was particularly convinced by our arguments in *Emergence* that this was a pseudofossil. Alas, as is often the case when dealing with a small number of enigmatic specimens, these arguments proved wrong.

Before proceeding further, I would like to mention the dangers of reversing oneself in science. I have had to reverse my earlier interpretation of this specimen not once, but twice in print. Not only is this painfully embarrassing, but it can do damage to one's professional reputation, a most valuable thing for a scientist.

As Peter Medawar once put it, science is the "art of the soluble," and the whole point of science is to come up with the correct solutions. There is a verifiable objectivity to the process of science unmatched in nonscientific fields. When the reward structures in science are functioning well, the greatest rewards go to the scientists who make the most important discoveries, the scientists who make the most correct interpretations, and those who solve the most important and difficult problems *correctly*. As a subscriber to this notion of science, I strive to get things right in my scientific work. Sometimes, in the face of new information, this may mean changing my stance on some issue. All too often, this process is skewed by political factors or by reluctance to change one's mind.

If one trusts one's own judgment, and then one feels compelled to change one's mind repeatedly about an issue, this may be an indication that the issue at hand is one of considerable importance (if not, better stop trusting one's own judgment). But intellectual caution is not always what is needed. As Daniel Dennett once said, we "often learn more from bold mistakes than from cautious equivocation." In my opinion, as a paleontologist one is not doing one's job unless one can occasionally be shown to be wrong. This is actually a healthy sign, for it indicates that the work is proceeding well within the realm of testable hypothesis and verifiable (or falsifiable) prediction.

My discovery in 1995 of Ediacaran fossils below the Clemente oolite of the Clemente Formation (the oldest convincing Ediacaran fossils known; see chapter 9) led me to reevaluate my interpretation of the structures in figures 3.2 and 3.3. I remember now standing in the room in the Biogeology Clean Lab that held Preston Cloud's enormous reprint collection and asking him whether my specimens were *Vermiforma*, a fossil he had described a few years before from the North Carolina slate belt.[7] He was noncommittal but reiterated that *Vermiforma* was a body fossil.

I recently returned to Cloud, Wright, and Glover's 1976 paper on *Vermiforma*. The fossils were found in Proterozoic strata of North Carolina. As part of this article, Cloud had published the description of a new species, *Vermiforma antiqua*. The journal *American Scientist* is a very unusual place to publish a description of a new species. *American Scientist* articles are generally summaries of already published research advances, not the original research itself. Cloud perhaps felt that such an unusual fossil merited an unusual publication venue.

As per Cloud's original description, *Vermiforma* is represented by seven individual specimens and fragments of several other specimens on a bedding plane surface (figure 3.4). The fossils occur in a laminated, greenish, volcaniclastic sediment. Cloud, Wright, and Glover inferred the depositional environment of the fossils to be "rather deep water," although no conclusive paleobathymetric indicators were associated with the fossils. They inferred from zircon Pb-U isotope dates an age of "close to 620 m.y." for the fossils, but due to structural complexities of the rocks in which they occur, their age is not closely constrained. The age of the fossils falls somewhere between 555 and 680 million years.[8]

Cloud interpreted *Vermiforma antiqua* as the body fossil of an elongate metazoan with possible annelid worm affinities. Cloud's figure 3 (repro-

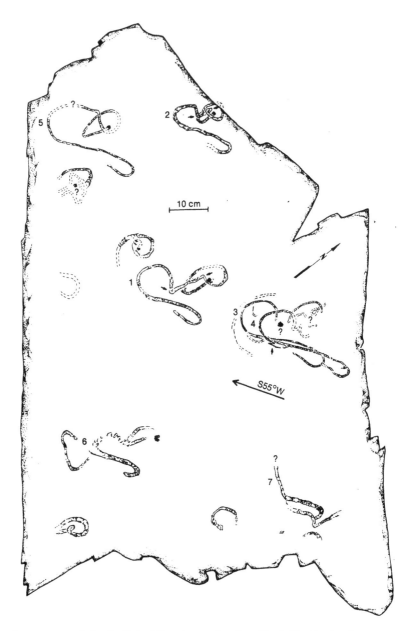

FIGURE 3.4: A sketch of a bedding plane surface from North Carolina bearing numerous specimens of *Vermiforma*. Note quasiholographic repetition of the shapes of the traces.

From P. Cloud, J. Wright, and L. Glover, III, "Traces of Animal Life from 620 Million Year Old Rocks in North Carolina," *American Scientist* 64 (1976):396–406.

duced here as figure 3.4) shows what he considered the "striking preferred orientation" of the structures. Cloud did not explain this preferred orientation of the structures from the perspective of his body fossil model. Nevertheless, Cloud's body fossil interpretation was accepted without reservation until 1992 by paleontologists who had considered the matter.

A close examination of the preferred orientation of the structures seen in figure 3.4 falsifies Cloud's body fossil hypothesis. The fossils show topological congruence even with regard to slight bends and curves in their sinuous forms. This degree of matching would be impossible for looping body fossils, even if they had been oriented by current. Cloud's small paired arrows in his numbered specimens 1 and 2 (figure 3.4) indicate two of these points of topological congruence, and similar points of congruence matching the spots indicated by the small left-pointing arrows can be identified in specimens 4, 5, and the fragmentary specimen immediately below specimen 6, and a point of congruence matching the target of the small right-pointing arrows can be seen in specimen 4. Not only do the longest stretches of each specimen line up with the compass direction S55°W, but nearly all sections of all specimens or partial specimens can be matched, with varying but generally very high degrees of precision, to equivalent sections on all associated specimens.

This quasiholographic repetition, in addition to overturning the body fossil interpretation, has not been described elsewhere in the fossil or sedimentary record and thus requires a special explanation, lest someone argue that the creatures were reading each others' minds.[9]

After my own reinterpretation of *Vermiforma*, I learned that Runnegar and Fedonkin had already reinterpreted it as a trace fossil. Crimes agrees with them in this assessment, as do I. Here is the current interpretation of both *Vermiforma* from both North Carolina and the Clemente Formation:

Vermiforma represents trace fossils, tracks formed in the direction of prevailing currents, as the tracemaker laid down a sticky but yielding mucopolysaccharide trail. The arcuate (roughly C-shaped[10]) impressions are concave in the down-current direction and are composed of fine sediment particles. These particles form a ridge with some relief. The ridges and associated depressions apparently formed as "heel markings," indicative of upward pressure on the leading or down-current edge of the adhesive holdfast, resulting from the fulcrum effect of water pressure on the part of the organism that projected into the water column (figure 3.5). At intervals, marked by successive barchans, the adhesive holdfast was loosened from its former position, was glided by force of current a short distance in a down-current direction on a sticky mucous film, and was then anchored into place, in part by hardening of the mucous cement and in part by lodging of the front of the holdfast as it slid forward and down the last barchan. With each successive translocation of the holdfast, the organism could rotate to give itself a preferred orientation with respect to the prevailing current. Thus, as the maker of the trace added the next increment of mucous trail, it was able to maintain a preferred orientation vis-à-vis the water current by rotating about a new, sticky but pliable layer of mucus, move downstream in a controlled fashion, and subsequently cement together sediment particles on its downstream end to form the next barchan (figure 3.5).

With each increment of trail growth, the last barchan was not obliterated by continued elongation of the slime trail. Instead, the organism glided over the old barchan, forming a nested, new barchan immediately downstream. The old barchan is preserved, possibly because the sediment particles of which it is composed are impregnated by the same viscous, hardening substance that forms the trail itself. Indeed, preservation of penultimate and earlier barchans is enhanced by the fact that they were covered by successive layers of slime. Each barchan received several coats, the number of layers being a function of the distance cov-

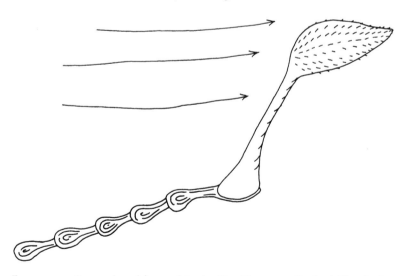

FIGURE 3.5: Proposed model to explain the *Vermiforma* trace fossils. A filter-feeding organism was glued to the sediment surface by a sticky mucus. At times when the mucus was soft, and stiff water currents (arrows) were present, the organism slid along its base, leaving behind a hardened mucous trail.

ered by each successive translocation of the organism. This distance in turn would be a function of the curing time of the mucous "bioepoxy." Larger secretions of mucus would result in longer glide translocations.

It is no wonder that Cloud misinterpreted *Vermiforma* as a body fossil. With its multiple layers of slime, it preserves in much the same fashion as a body fossil. Mucus secreted by modern marine invertebrates has these types of properties.

The mucous holdfast acted as the underpin for a part of the organism that, because of its shape, had an inherent tendency to orient itself in a current. This current-obstructing object is presumed here to have functioned as the organism's filter-feeding device. The shape of this structure is unknown; one possibility would be a teardrop-shaped organ, covered with sticky mucus to trap food particles. Any object with this shape would have an inherent tendency to orient itself with the point of the teardrop pointing away from the current. Alternatively, the structure may have been parasol shaped, with a food-catching sieve on the downstream side.

Gliding on a slime track might seem to be an inefficient way to orient oneself in a current; indeed, such a strategy would seem unnecessary for an organism with muscular (e.g., sea anemones, worms) or skeletal (e.g.,

FIGURE 3.6: *Aspidella*, an ovoid Ediacaran fossil from eastern North America. Greatest diameter of individual fossils 1 cm.

sea lilies) support. It is thus possible that the *Vermiforma* tracemaker was a large, mucus-forming organism but was not an animal. Its trace fossil shows no evidence of muscular activity; the preferred orientation of the creature was a passive affair requiring no muscular contractions. Indeed, this glide-and-glue strategy would permit a firm attachment site and allow periodic readjustment in a creature without muscles.

The *Vermiforma* creature could have been a large protist. Modern analogs exist for centimeter-sized, holdfast-attached protists that extend upward into the water column.[11] Copious mucus secretion is no metazoan monopoly; Proterozoic cyanobacteria used periodic, crescentic secretions of multiple thick mucus layers as a means of locomotion by gliding.[12]

Judging from the shape of the mucus-bound barchans, the holdfast that formed *Vermiforma* was flat on its underside and, at least on its leading edge, roughly elliptical or perhaps parabolic (figure 3.6) as viewed from above.[13]

After coming up with this new interpretation of these Mexican *Vermiforma* fossils, I decided to revisit associated specimens from the Clemente Formation of the Cerros de la Ciénega that, in my dissertation, I had described as sand volcanoes. It did not take me long to decide on a new interpretation for these as well. After renewed study of the specimens, I noticed faint cross-stratification[14] that I had missed before on the side of one of the specimens. The orientation of these cross-beds showed that the orientation of the pitted mounds must be convex *downward* rather than convex upward, in which case they could not possibly be sand volcanoes as per my earlier interpretation. The central dimple on the mounds connects to what I had thought was the cylindrical vent for the fluidized sand. This must now be some sort of upright structure, perhaps the holdfast for a frond of some sort.

Concentric wrinkles mark the edges of some of the mounds. Warts of sediment (which I now interpret as solidified clots of mucus-bound sand) occur on the flanks. These inverted pit-and-mound structures are clearly stationary versions of *Vermiforma*, in which the holdfast stayed put rather than traveling slowly downstream in the direction of prevailing current.

Interestingly, Seilacher has described a somewhat similar form, *Vendospica diplograptiformis*, housed in the Schiele Museum of Natural History, Gastonia.[15] Like *Vermiforma*, this form is from Proterozoic strata of North Carolina. Seilacher interprets these markings as drag marks left by the down-current transport of stiff bodies with pointed basal tips. The organism creating the tracks is unknown.

The detachment and readhesion of the mucus-generating holdfast of the *Vermiforma*-former allowed the organism, via periodic course correction, to maintain a preferred orientation with respect to prevailing currents. The elongate trace formed by the mucus-generating holdfast of the *Vermiforma* tracemaker can thus be read as an incremental record of change in position with the direction of currents.[16]

The evidence of current-influenced orientation strongly suggests filter feeding as the nutritional mode for the *Vermiforma* organism. Perhaps the same sticky mucus that was used for gliding was also used for trapping food particles from the water column. The motion of the cohort of *Vermiforma* individuals on the same bedding plane was apparently synchronized. This is not an unreasonable possibility, considering the recently published observations of Gooday, Bett, and Pratt showing that growth phases of the xenophyophore protist *Reticulammina labyrinthica* were synchronized in different individuals.[17]

Seilacher has long had suspicions about *Vermiforma*, as indicated by our conversation on the subject in Namibia in August 1993. He saw it as a useless taxon that would be bouncing around in the literature because "for lack of characteristics you cannot kill it." I think Seilacher's judgment is unnecessarily harsh, especially considering his own description of *Vendospica diplograptiformis*.

This is one of the few times I have found myself in direct disagreement with Seilacher. A lecture I gave in 1995 at Yale was well received, but Seilacher had questions about my interpretation of Cloud's *Vermiforma*. Seilacher felt it could be a pseudofossil, formed by already deposited beds sliding past one another, as seen in the offset mudcrack slab from Namibia (color plate 3). I replied that that would be a permissible interpretation, except that the traces on the North Carolina bed are not exactly congruent, but rather are quasiholographically congruent. This arrangement would be impossible under the sliding bed interpretation, in which they would need to be exactly similar unless the bed was stretching, in which case there would be a gradational variation in the markings from one end of the slab to the other.

Look carefully at the sketch of the slab. Track M has a proportionally longer long stretch than does Track N. This rules out Seilacher's interpretation of pseudofossils formed by bed gliding. Tectonic deformation of the bed is also ruled out because no conceivable type of stretch deformation can account for the bedding pattern. However, the pattern is nicely explained by minor variations in organism gliding speed.

Notes

1. M. A. S. McMenamin and D. L. S. McMenamin, *The Emergence of Animals: The Cambrian Breakthrough* (New York: Columbia University Press, 1990).

2. Gneisses are foliated metamorphic rocks.

3. A migmatite is a metamorphic rock that has been heated to such an extent that it becomes difficult to distinguish from an igneous rock.

4. A field term for dark, fine-grained igneous rocks with a greenish tint.

5. M. A. S. McMenamin, S. M. Awramik, and J. H. Stewart, "Precambrian-Cambrian Transition Problem in Western North America: Part II. Early Cambrian Skeletonized Fauna and Associated Fossils from Sonora, Mexico," *Geology* 11 (1983):227–230.

6. A metazoan is any multicellular animal with organ systems and with cells arranged in two layers in its embryonic stages.

7. P. Cloud, J. Wright, and L. Glover, "Traces of Animal Life from 620 Million Year Old Rocks in North Carolina," *American Scientist* 64 (1976):396–406.

8. See pp. 396–397 in P. Cloud, J. Wright, and L. Glover, "Traces of Animal Life from 620 Million Year Old Rocks in North Carolina," *American Scientist* 64 (1976):396–406.

9. A pseudofossil interpretation for the *Vermiforma* structures, such as grooves formed by one bed gliding over the other, can also be ruled out because the *lengths* of the structures differ, an impossible situation under the gliding bed hypothesis.

10. Like a barchan sand dune.

11. T. E. Delaca, D. M. Karl, and J. H. Lipps, "Direct Use of Dissolved Organic Carbon by Agglutinated Benthic Foraminifera," *Nature* 289 (1981):287–289.

12. S. Golubic and H. J. Hofmann, "Comparison of Holocene and Mid-Precambrian Entophysalidaceae (Cyanophyta) in Stromatolitic Algal Mats: Cell Division and Degradation," *Journal of Paleontology* 50 (1976):1074–1082.

13. The Precambrian fossil *Aspidella* has approximately the same simple oval shape, and may represent a nongliding version of *Vermiforma*.

14. Formed by ripple marks in ancient sediment, cross-stratification gives geologists clues to both the direction in which the water was moving the sediment (thus forming the ripples) and the relative position of the top and bottom of the bed in which the ripple marks formed. Structures that can distinguish the tops from the bottoms of beds are called geopetal structures.

15. See p. 397, figure 6 in A. Seilacher, "Early Multicellular Life: Late Proterozoic Fossils and the Cambrian Explosion," in S. Bengtson, ed., *Early Life on Earth*, pp. 389–400 (New York: Columbia University Press, 1994).

16. *Vermiforma* is thus a new type of geopetal structure (in other words, it can be used to tell what are the tops of sedimentary beds) as well as a sensitive indicator and recorder of fluctuation in the direction of ancient seafloor currents.

17. A. J. Gooday, B. J. Bett, and D. N. Pratt, "Direct Observation of Episodic Growth in an Abyssal Xenophyophore (Protista)," *Deep-Sea Research I* 40 (1993):2131–2143.

4 · The Nama Group

[Outrageous] hypotheses arouse interest, invite attack, and thus serve useful fermentative purposes in the advancement of geology.

—John K. Wright[1]

In spring 1993 I received an invitation from Professor Adolf Seilacher (figure 4.1) to join his field party for an expedition to Namibia. Seilacher[2] had just been awarded the prestigious Crafoord Prize by the Swedish National Academy of Sciences and was wasting no time in putting the prize money to good use. I was honored by his invitation and hastened, on April 18, to accept: "I would be very pleased to join you on your trip to Namibia this summer, and I will be happy to serve as raconteur."

Seilacher is generally regarded as the best paleontologist of his generation, a gifted teacher, an extroverted lecturer, and an observer of nature with an awesome ability to make key observations. His Vendobiont theory, in which he proposed a radical reinterpretation of early fossils called Ediacarans, is not the first, nor perhaps even the most important instance in which Professor Seilacher has changed our view of life. In his 1972 paper "Divaricate Patterns in Pelecypod Shells," following a line of paleontological thought begun by George Gaylord Simpson, he claimed that divaricate color patterns and other conspicuous traits of organisms are in many cases nonfunctional (read nonadaptive). First greeted with skepticism, this paper now holds a very special position in paleontology. In his own words, "The basic form [of divaricate patterns] is assumed to be less controlled by adaptation and phylogeny than by a common principle of shell growth."[3]

Seilacher's recognition of fabricational, nonadaptive characteristics has dramatically enhanced the sophistication with which we can understand evolution. It leads to the most crucial and difficult counterintuitive lesson of evolutionary theory: the idea that self-organizing morphogenetic mechanisms are the processors of evolution. In other words, highly complex forms of life can appear for, at least from the point of view of an adaptationist, no reason at all. (An adaptationist is one who views all morphological traits of organisms as having an adaptive, or "survival," value.)

FIGURE 4.1: Dolf Seilacher in the field in Namibia.

Seilacher focuses on the process of morphological incarnation. This focus pervades Seilacher's professional work. As described in this chapter, a huge slab we had excavated was successfully reproduced as a silicon peel. I watched Seilacher walk away from the prize specimen, leaving it in the field. Satisfied that the artificial cast will provide the information necessary for interpretation of the process of growth and life of these forms, Seilacher was content to leave the choice specimens in Namibia.

This focus on process extends to the illustrations in Seilacher's publications. He insists on producing his own illustrations because drawing enforces careful observation. Each hand-drawn illustration in a Seilacher paper contains multiple images. In my opinion his illustrated articles are like fractals: The illustrations are like manuscripts in themselves and can be read as such.

I greatly looked forward to joining Seilacher in Namibia. In a heady mood, I began my first entries for the trip:

July 25, 1993: Gate 4, Springfield, Massachusetts Bus Station.[4] Corporate headquarters since 1933.

Seilacher first entered the United States in 1954 via El Salvador and Mexicali. He met the border guard on foot. Later he earned a $100-per-month stipend at Stanford University. He remembers Professor A. Myra Keen trying to teach students how to properly pack specimens. Boxes were dropped to demonstrate the efficacy of various packing strategies. Failed attempts were not graded on a curve.

I'm still standing in the Peter Pan bus station on a sultry New England Sunday afternoon. Ambulances and towing company vehicles are preparing for a parade. Passenger cars as well, with the Puerto Rico flag flying above both front fenders.

A Spanish-speaking lady is ahead of me in the ticket queue, with her son and her henna-colored hair. My ticket to Logan airport costs $21.00, with a bus change in Boston, so the driver tells me. We stand in line waiting to board the bus, and a white woman ahead blows a cloud of smoke my way and then crushes out her cigarette on the concrete. The bus fills two-thirds full with Hispanics, blacks, whites. Two black men board with white baseball caps and purple visors. One has a Walkman deployed while he impatiently chews gum. The other man is traveling with his son, who has a fashionable bowl cut. The father wears white "home boy" long shorts and shirt, and is playing a Nintendo "game boy."

Peter Pan Bus Lines names its buses: "Peter's Kiss," "Captain James Hook." I cannot read our bus's name while seated; on a whim I ask the driver for the name and with annoyance he replies "I have no idea." He pulls us out of Gate 4 and Nintendo spills his soft drink. A long tongue of soda runs down the aisle. Ignoring the heat, a driver in a gorilla suit passes in a jeep, flying Puerto Rico colors.

A water tower dominates the landscape where Interstate 91 turns off onto the Massachusetts turnpike. 2:30 P.M. we stop in Worcester. The Irish influence is evident in the names of roads: Brosnihan, Kelly, Kelley. Still curious about the bus's name, I step out to have a look. "Tick! Tick! Tick!"

The Coney Island Hot Dog restaurant, with giant yellow mustard dripping from its hot dog marquee, abuts our parking lot. Parking in this lot is not free; motorists must deposit money in a yellow "Pay Here" box; the box is a micro-cosm of the parking lot, and the money is deposited in the numbered slots:

Deposit money in same number slot as your parking space. Fully insert dollar bills one at a time, folded three times, then insert coins.

A gardener has parked here. He has a blue bandanna over his hair and a Z^{71} long-cab pickup truck. His two daughters are with him, one has gloves, and both are here to help. Dad pulls a lawnmower out of the back of the pickup.

We resume our journey through Worcester. Kelley Square Liquors. Wein-traub's Delicatessen. Cloverleaf and Star of David. Diamond Inn. St. Stephen's Catholic Church. Grafton Street. Pilgrim Street Not a thru street. Puritan Street Not a thru street. Grafton Street has elements of the Americana highway strip, but is more classy, like Cape Cod. Sign (also encountered on the Cape): "Thickly Settled." Route 122, south. Purple loose-strife grows wild near the Turnpike entrance. It likes the wet, low areas along the Pike. Reduced Salt Area. One if By Land Bus Tours. Framingham.

3:33: Boston, South Station. Everything Yogurt. James Cook and Company Live Lobsters. Callahan Tunnel. Logan Airport. Long Wharf Marriott.

Boston, with evenly spaced trees and buildings, is lovely on this July day as seen across the harbor from Boarding Gate 6. Seagulls cross the water with strobelike wing beats. The Lufthansa symbol is a crested bird in flight, but the bird is very slender, as if an eagle of the Third Reich had been plucked. Airline clientele seems mostly white. A German woman carries a travel bag entitled, across its side in large letters, "The Ultimate Solution." We will be flying a DC10–30 Bremerhaven. An Aer Lingus E1-ASJ stands near our Lufthansa jet. Lufthansa jets have an interesting comma-shaped pattern in blue paint on the jet cones reminiscent of the spirals Nazis painted [in red] on the nosecones of the Messerschmitt ME-109.

5:05: We're still waiting to board. The German tourist sitting in front of me with blue baseball cap with a "G" and a blue tee shirt with "Faith No More" printed on it in numerous languages. But the other languages are transliterations, not translations.

A curt head nod seems to be de rigueur when German adults meet. On board, Flight Steward H.-D. Goetz reminds me of Adolf Seilacher, the man responsible for my flight to Germany.

Seilacher has invited me to join him on a paleontological expedition to the former German colony of South West Africa, now the newly minted country Namibia. With his team of German scientists and technicians, we will search for half-billion-year-old fossils belonging to what paleontologists call the Ediacaran biota. German scientists discovered the first Ediacaran fossils in South West Africa in 1908. Most paleontologists believe that the members of the Ediacaran biota represent fossils of the earliest animals. Seilacher is notorious for his disagreement with this conventional view.

July 26, 1993: Airport Frankfurt Sheraton, Frankfurt, Germany.

I am seated in the lobby of the Frankfurt Sheraton, outside the Hiller shop. Artwork on the wall looks like a cross between a nebula and a poached egg. The Frankfurt Airport Center is encased with attractive slabs of Rapakivi Granite. Rapakivi granite is a type of igneous rock that when slabbed and polished appears peppery from a distance. It is a popular material for überclass coffee tables. I imagine that much of a batholith[5] must have been exhumed in order to provide enough Rapakivi to enclose such a large building.

Back in the airport, an impromptu air museum adorns the B departures area. A red Fokker D7, a Fokker with speckled camouflage, a replica of Lindbergh's Spirit of Saint Louis (engine by Ryan), a red Morane MA 317 (1933). A copy of Richthofen's red 1917 Fokker Dr. 1 triplane prominently displayed. Three wings makes a plane slow but very maneuverable.

As in the time of Hitler, Germany is still a nation of fliers. The display continues into a pictorial history of the Frankfurt airport, 1909 to the present. The 1911 panel shows a hot air balloon; by 1912 a giant propeller marked the airfield. Early in the display is an astonishing photograph of a First World War dogfight; 10 or more biplanes gyrate in an aerial brownian motion of battle. A German Albatross with the Iron Cross dives improbably.

The history of the Zeppelin comes next. Zeppelin mementos (medals, toys, guidebooks, china) visible in plexiglass cases. The 1932 panel shows a Dornier "Jumbo" float plane with six pairs of propeller engines, with the propellers facing opposite directions in each pair. The huge plane is surrounded by kayakers.

Landung der Dornier DO X auf dem Main 1932 "Jumbo."

The display then moves to several panels of the Nazi era, 1936–1945. Light vandal scratches across the swastika on a flag in one of the photos. The 1936 panel shows a plane that may have been an early version of the Stuka, but perhaps out of a sense of national shame, none of the Nazi era panels show the famous planes of the Luftwaffe, not even the famous Messerschmitt ME-109 or the early Luftwaffe jets. (The Messerschmitt ME-109 does, however, appear in a nearby airport toy shop. The Spanish Air Force used the plane for years after World War II.) The 1945 panel shows the eagle of the Third Reich being removed from its pedestal.

The rest of the panels emphasize postwar peaceful flight, and the 1952 panel emphasized cosmopolitan Germany. Bilingual signs: "Uskunft-Information." A Volkswagen beetle with a sign on its rear, in English "Follow Me." The crew peering through the glass of a Pan American "Clipper Climax."

I had a long conversation with a young woman from near Hannover on the transatlantic flight. She spent the year in Boston, and did not want to leave the States to return to Germany. Her father drives buses on the Autobahn. She was in Tunisia when the Berlin Wall fell, and she believes that the two Germanys should have remained separate countries.

I walked past a symbol for Church services; it resembles the cross insignia on the Fokker D-7. I stepped outside, and walked along on a ramp footpath going nowhere in particular. The air was cool and inviting. The dense trees across the Autobahn are very unlike those in New England; rather than tapering tannenbaum style they have the vegetation concentrated [and flattening out] near the top, giving them an African, Serengeti look.

Returning to the airport, I stepped into Namibia. I was pleased to find a second airport gallery with a special July-August botanical exhibit [of plants from Namibia] on loan from the Frankfurt Botanical Garden (Stadt Frankfurt am Main Palmengarten).

As you can see from the notes above I was trying hard to hone my observational skills in preparation for Africa, hoping to improve my abilities as field rapporteur.

Before proceeding I want to make a comment about what I have written here. You will find examples of racial tension in what follows. I have tried to be sensitive and respectful to all while faithfully reporting what I saw and heard. My own feeling on the matter is that although we are all members of the same species, *Homo sapiens,* there are indeed profound racial differences. No one fully understands the historical or scientific meaning of these racial differences, and that is why no one knows what to do about them today. All we know (and even this can be questioned) is that it all started in Africa. The great Namibian geologist Henno Martin considered the matter of racial differences and concluded that anyone who understands Mendelian genetics should realize that human racial differences are unimportant.[6]

The airport exhibit had what appeared to be a comprehensive display of Namibian succulents. There was the living stone plant *Lithops gracilidelineata* (family Aizoaceae) from the Namib desert. *Lithops,* limited to two swollen leaves, resembles a persistent seedling (color plate 4), although old leaves do dry up and new ones develop inside and emerge in late spring. The Thompson & Morgan seed company calls *Lithops* "a perfect example of nature's adaptability. Neat little plants that look for all the world like stones. They are Succulents, easy to grow and create a great deal of interest."[7]

Also present in the exhibit is the finger-shaped *Fenestraria aurantiaca* (also family Aizoaceae) from Kap, Kleines (Lesser) Namaland, Namibia. The spiky-leafed *Aloe erinacea,* the spotted-leaf *Aloe hereoensis* v. *h.* of Damaraland, and the astonishing-looking aloe tree *Aloe dichotoma* (kokerbaum) and *Aloe ramosissma.* All the aloes are family Liliceae, from Großes (Greater) Namaland, which must be the center of aloe diversification.

The cactus *Euphorbia virosa* (family Euphorbiaceae) occurs in Namibia (Lesser Namaland). *Pachypodium namaquanum* of family Apocynaceae (Kap, Greater Namaland) has an awkwardly swollen trunk. The *Cyphostemma* species (family Vitaceae), including *C. seitziana* (Grandilla Mountains), *C. juttae* v. *juttae* (Greater Namaland; larger and treelike), and *C. currari* (Kap, Elephant's Bay), all have thick trunks and lobate leaves.

Finally, I saw a live specimen of the odd plant that is emblematic of the botanical curiosities of Namibia: *Welwitschia mirabilis* (family Welwitschiaceae). *Welwitschia* is one of the few surviving members of a group of plants known as the Gnetales, a botanical order with only 71 species that includes *Ephedra* (Mormon tea) and *Welwitschia mirabilis.*

Welwitschia looks like a low, wilted head of romaine lettuce grafted to a giant parsnip, and is indeed a curious-looking plant. Its thick, inverted conical stem can reach up to a meter and a half in diameter. The plant is broader than it is tall. Like *Lithops,* the mature plants have only two leaves, leading some botanists to describe the plant as a persistent seedling, although it does have a complete reproductive cycle. Even here there is a unique feature, however. Unlike any other plant, during fertilization a tube grows upward from the egg to unite with the pollen tube.

The last plant I encountered in the exhibit was *Lithops erniana*. This living rock is named for the Erni family, members of whom I was to visit in Namibia. I was struck in this exhibit by the large number of unusual plants, although many of them were familiar because they are grown as ornamentals in southern California, where I attended school. The Nama is a very ancient desert; perhaps this is why it has accumulated such a variety of unusual-looking plants.[8]

Leaving the grow lights of the desert plant display, I wandered over to the other side of the air exhibit gallery. The Messerschmitt one-man jet ME-163 "Komet" is the only World War II–era plane. This seemed odd, even considering the space limitations in the airport lobby, because the war produced an astonishing diversity of German aircraft, including formidable jet fighters that were design precursors to some of the modern U.S. military aircraft. The ME-163 was not one of these, however. It was a tiny delta-winged plane used to protect synthetic fuel plants against Allied bombers, and could stay aloft only for about 7.5 minutes. Its insignia is a flea headed skyward with a rocket jet coming out of its tail:

Wie ein Aber Floh Oho! [Small like a flea, but deadly!]

The shape of the ME-163 is echoed in a suspended bitail AV36 tiny red and white glider. Seilacher has recounted how, growing up in Nazi Germany, he and other young men were encouraged to fly gliders. No enclosed cockpit, just an open air wooden seat and a rudder control stick. A nation of flyers.

Further wandering brought me to a display on the new nation of Namibia. Formerly it was the international territory of Namibia, which was once the German colonial territory of German South West Africa. It is situated on the Atlantic coast of southwestern Africa, bounded on the south by South Africa, on the east by Botswana, and on the north by Angola. Its area is 824,290 km, but it has a population of only 1.4 million, and many of these people live in the capital city Windhoek.[9] Diamonds are the most important export, but copper, lead, zinc, and cattle are also exported.

Like most African nations, Namibia has had a troubled history in the twentieth century. However, Namibia's story has a happy ending for now. South Africa took control of South West Africa as per a League of Nations mandate on December 17, 1920. On dissolution of the League of Nations in 1946, the United Nations inherited its supervisory authority for South West Africa and in the same year refused South Africa's request to annex South West Africa. South Africa retaliated by refusing to release the territory to a United Nations trusteeship.

The International Court of Justice ruled in 1950 that South Africa could not unilaterally annex or modify the international status of the Namibian territory. A UN resolution in 1966 declared the 1920 mandate void and terminated. South Africa rejected this declaration, and the status of the region remained in limbo.

The UN General Assembly voted to rename the territory Namibia in 1968. The International Court of Justice was heard from again in 1971, this time ruling that South Africa's presence in Namibia was illegal. An interim government was established in 1977, and independence was to be declared on December 31, 1978, but the resolution was rejected by the major UN powers. Elections supervised by the United Nations in April 1978, under a South Africa–approved plan, led to a political abstention by the militant South West Africa People's Organization (SWAPO). A Multi-Party Conference (MPC) held successful talks with SWAPO, petitioned South Africa for self-government, and on June 17, 1985, installed the Transitional Government of National Unity. Negotiations held in 1988 led to withdrawal of Cuban troops from Angola and South African troops from Namibia.

The Transitional Government resigned in February 1988, and Namibia finally achieved its independence on March 21, 1990. Namibians are thankful for this quiet triumph of diplomacy: Billboards in Windhoek proclaim, "Thank You United Nations!" However, South Africa still controls the territory surrounding the major port, Walvis Bay, and the lucrative diamond trade is largely in foreign hands. The Caprivi Strip panhandle, a thin belt of Namibian territory, extends due east from the northeast corner of the main area of the country, as a buffer zone between Angola and Botswana. The first Namibian coins appeared in 1993.

The first president of Namibia is Sam Najoma, a black man of pleasant visage with a broad smile. Namibia bills itself as the smile of Africa. The national anthem is "Namibia Land of the Brave" (music text, Axali Doeseb; arrangement, Konrad Schwieger). The national coat of arms is a green shield with a red stripe and yellow sun, surmounted by a diamond headband and an African fish eagle, and with an oryx on either side. Underneath the shield is a sprawling *Welwitschia* plant, which the display notes is a "fighter for survival and is therefore a symbol of our nation's fortitude and tenacity." Below *Welwitschia* is the motto "Unity, Liberty, Justice," the latter perhaps a nod to the International Court of Justice.

Back to notebook:

> Colonial troops look like jackbooted German cowboys.
>
> Tribal art. Some of the most interesting woodcarvings of animals emerging from a dark manzanitalike wood. Weathered wood surface blends with burnished necks of swanlike bird and antelope gracefully stoops to drink. Following natural wood contours leads to grotesque shapes.
>
> Flight LH 568 to Windhoek, then Johannesburg. Mostly whites on flight. Black family with crying baby sent back to check through stroller; two big glances were exchanged that could cut through tungsten. After 15 hours waiting for flight, I'm ready. Germans love to smoke. More tobacco stores than fast food. Even the USO (50 years 1942–1992) is sponsored by Marlboro. Talk between airman and army man over what the Germans will inherit (new medical facilities) as the Cold War bases close. Where did they get that fried chicken?
>
> Keine Umstellung! No jet lag.

Several possible themes emerged as I waited for Seilacher to pick me up at the Windhoek airport. Plants exposed to rigors of the desert develop into weird succulents. Luftwaffe aircraft develop under the rigors of war into jets, rockets, and eventually NASA. Ediacaran organisms under the rigors of predation give way to shelly trilobites. Nama natives develop under the rigors of colonialism into the Republic of Namibia. The Bushman race, called *The Harmless People*[10] and pressed into slavery by the Ovambo tribe, have had a sorry recent history. Could they have a statesmanlike future in the new Namibia?

Shortly after my Air Namibia (a converted Lufthansa jet) flight landed, Seilacher and company picked me up at the Windhoek (Eros) airport in a Volkswagen minivan. The driver sat on the right, a legacy of British takeover. I had a lunch of gemsbok (oryx) at a restaurant on a Windhoek terrace called the Gathemann. We spent the night in the lovely Motel Safari in Windhoek. As night fell, I saw the Southern Cross for the first time.

Over a beer in the Motel Safari, Seilacher explained his rationale for bringing us all together in Namibia.

Seilacher confided to me that he hoped we would have informative discussions with Bruce Runnegar and Jim Gehling over beer in Aus in the evenings, and would not be too tired to engage in fruitful discussion. He also expressed hope that Bruce and Jim would not be converted to his way of thinking because it would destroy the whole discussion. Seilacher's main aim in this exercise was to find out how far a reductionistic (as opposed to a holistic) way of thinking could go in terms of interpreting Ediacaran fossils using organisms and sediment types known today. In other words, Seilacher planned to stand firm with his unorthodox arguments and he wanted to see how far his adversaries could get with their traditional actualistic arguments in which the present is viewed as the key to the past. As Seilacher put it, in his counter-actualistic vein, "Of course physical laws were the same but habitats were not the same because . . . the biology was different."

Leaving the next morning at 8:39, we stopped in at the Namibian Geologic Survey on the corner of Robert Mugabe and Lazarette Streets. The Survey building is attractive, with a red corrugated roof and yellow with green trim that matches the Namibian flag flying overhead. The National Monuments Council of Namibia is across the street.

Windhoek seems preternaturally neat and tidy, a result of civic pride, inexpensive labor, and new buildings. Men in fatigues and machine guns are seen on street corners, however.

At the Survey we saw an unusual case of shifted mudcracks, looking like what a numismatist would call a double strike (color plate 3). First described in 1975 by R. M. Miller in the *Journal of Sedimentary Petrology*,[11] these unusual mudcrack features showed a distinct displacement between identical polygonal patterns, and the specimens in the Survey courtyard captivated Seilacher's attention. We spoke with Survey geologist Charles Hofmann about other unusual sedimentary structures.

The main road south out of Windhoek, toward Rehoboth, is B1. As we headed onto the Main Road, we passed the Tropic of Capricorn as a highway marker. Because of the paucity of long pieces of wood, the barbed wire fences are built with staggered short pieces (figure 4.2). Weaverbirds build solitary nests in clusters on single trees, or in giant communal nests.

Much of the land in this region is dedicated to the rearing of Karakul sheep. The wool of adult sheep is coarse and uncomfortable to the touch, but is nevertheless useful for heavily worn or abused items. It is so water repellent that airline seats made of the wool of Karakul sheep will float. The moth larvae of New England, who usually relish wool, refuse Karakul wool. A Namibian story, perhaps apocryphal, tells of farmers filling potholes with Karakul wool. On B1, the sheep are protected by special antijackal fences, dug into the ground so the jackals cannot burrow underneath.

Highway B1 goes all the way to Cape Town. Posted speeds are fast (100 km per hour) but the roads are narrow and two lane, not unlike

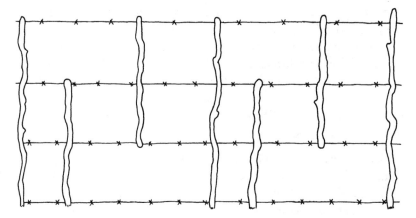

FIGURE 4.2: A Namibian-style barbed wire fence, which economizes on long fence-posts by using shorter ones.

the desert roads of Mexico. Every few dozen kilometers we saw telecommunication transmitting towers—white, monolithic landmarks.

Just south of Windhoek we encountered a troop of baboons. Near the Dabib River are interesting, mesalike rock formations. As we approached Mariental, we turned west toward Hardap Dam to visit a favored cafe, but it closed at 2:00 P.M.

Some of the bushes in this savanna land look like creosote bush, and a few of the trees look like Palo Verde of the American West, but the majority of the trees are unfamiliar to me.

Windmills for pumping water on farms are very common. There are some farms south of Hardap Dam, able to eke out a living in the plateau subdesert alkaline soils. Brown, loaf-shaped termite mounds are common where soil conditions permit. Gazing out the VW window, I saw an ostrich farm on the left. A strutting male was exhibiting his feathers to the females. In Mariental we saw the Drankwinkel (corner liquor store) and Makelaars (real estate agency).

During the drive discussion centered on the work of Edgar Dacqué (descended from Huguenots), a famed functional morphologist and biogeochemist.[12] Dacqué's 1921 book *Vergleichende Biologische Formenkunde* (*Comparative Biological Morphology*) appears to have had a formative influence on Seilacher's thinking. Foremost is the notion of biological synergism. For example, the interaction between the shell and the mantle of the mollusk's body leads to accretionary growth.

We reached the Quiver Tree or Kokerboomwood (*Aloe dichotoma*) Forest as nightfall approached. These ludicrous trees command attention, being unlike any other plant I have ever seen (color plate 5) except the Joshua tree (*Yucca brevifolia*) of the California desert.

The Joshua tree "is an amazing plant when thought of as a relative of the Lilly."[13] Found throughout the Mojave Desert of California, it can reach a height of 40 ft or more. This tree is responsible for what Robin Williams calls the Joshua tree principle.[14] Years ago she received a tree identification book for Christmas, and she decided to identify the trees in her neighborhood:

> Before I went out, I read through part of the book. The first tree in the book was the Joshua tree because it only took two clues to identify it. Now the Joshua tree is a really weird-looking tree and I looked at that picture and said to myself, "Oh, we don't have that kind of tree in Northern California. That is a weird-looking tree. I would know if I saw that tree, and I've never seen one before." So I took my book and went outside. My parents lived in a cul-de-sac of six homes. Four of those homes had Joshua trees in the front yard. I had lived in that house for thirteen years, and I had never seen a Joshua tree. I took a walk around the block, and there must have been a sale at the nursery when everyone was landscaping their new homes—at least 80 percent of the homes had Joshua trees in the front yards. *And I had never seen one before!* Once I was conscious of the tree, once I could name it, I saw it everywhere. Which is exactly my point. Once you can name something, you're conscious of it. You have power over it. You own it. You're in control.

Quiver trees are remarkably similar to Joshua trees with regard to the form of branches and trunk. Quiver trees or kokerboomwoods are aloe plants (like the Joshua tree, also a lily relative) that grow straight up at first and then dichotomously branch to form a fairly full, spiky crown. This must represent an independent evolution or development of the weird tree habit. These plants approximate what would result if one were to saw off all of the leaf-bearing branches from a tall magnolia tree, and then stick aloe house plants to the sawed-off ends of the limbs.

The swollen trunk bases of quiver trees are covered in platelike scales. Each plate has concentric growth lines, making it look similar to the plates on the shell of a turtle. Young branches have smooth bark that splits with age; the splits later heal over as bark. Plates are bordered by old splits.

The wood of the quiver tree begins in the center with a fibrous core, surrounded by a tough cylindrical layer called the quiver, or inner core. Bushmen hollow out this inner core to make quivers for their arrows.

The inner core is surrounded by a fibrous cortex, which in turn is surrounded by a bark generated by a unifacial cambium (color plate 5).

We spent the night of July 29 in the Travel Inn of Keetmanshoop. The strains of the Namibian version of country music were audible, and not quite as twangy as the American version. "Fill up and feel good," promises the local Shell gasoline station. We stopped in at the local Namibian post office. Bank of Namibia money would be available in September. One denomination was to feature Chief Hendrik Witbooi (1840–1905), a black freedom fighter with a pinched face, a wide-brimmed hat, and a rifle. The tourist center across the plaza from the post office has an Africa-shaped rock and desert rose gypsum crystals from Lüderitz. Seilacher told how he served in the navy in World War II on the *Weser* (named for a river in Northern Germany), a German minesweeper.

We arrived at 1:20 P.M. at Fish River Canyon, a cross between a badlands and the Grand Canyon. We stopped for rest at a visitor's lookout built of stonework overlooking Fish River Canyon. A tremendous angular unconformity[15] is visible on the far side of the canyon. Odd, squat grasshoppers inhabit the rim of the Canyon. They look like a cross between a pebble and a horned toad (which of course is a lizard). Peter Seilacher, son of the paleontologist and an artist himself, made sketches of the canyon. We found "antinodules" in the sedimentary rocks of the rim. These are 1- to 2-cm-diameter weathering structures that one could create if one were able to take small ice cream scoops out of the rock. They are probably a result of differential weathering of weakly cemented sediment. I picked up a specimen (1 of 7/29/93) of a possible Ediacaran fossil.

As the sun was setting, we drove into Aus (the name means "way out there") and took up our lodgings in the Bahnhof Hotel (color plate 6). Aus reminded me of the proverbial one-horse frontier town of the old American West. The lights go out at 10:30 P.M. The telephone at the hotel is a 13-party line. Rooms were cold at night because of an air vent that could be neither closed nor opened directly to the outside. No hot showers until Tuesday because all the gas cylinders were empty.

The Bahnhof Hotel is a one-star establishment, the single star proudly displayed, of skeleton keys, chipped paint, questionable ability to heat water when more than one room is using it, drafty windows, and no insulation during the Aus winter. The owner, Hans Theo Lubouski, was offering the hotel for sale for 300,000 rand. The managers, Frik and Bets Swanepoel, declined to purchase the property because it would require too much capital investment to repair.

The morning of July 30 we drove from the hotel to Aar Farm, owned by Hellmut Erni. The van traveled over a red dusty pad (desert jeep trail, pronounced "pat"). We met first with Erni's nephew-in-law, and then visited his farmhouse. After this we drove to the first locality for *Pteridinium* fossils.

Hellmut Erni is a former Swiss Air pilot (he retired on a pilot's pension). His nephew Wilfred works as a supervisor of road construction on the highway linking Keetmanshoop to Lüderitz. He has clear blue eyes and a weather-resistant face that evokes Heidi's Swiss grandfather. He spoke to us in German of how his father (Hellmut K. A. Erni, "Old Hellmut") had emigrated from Switzerland, originally planning in about 1911 to go to America, but after speaking to a friend also about to leave the country had decided to come to the South West Africa German colony. Hellmut Erni worked as a handyman and diamond prospector, eventually saving enough to buy the hotel in Aus, recently built by the Schutztruppe. When British and South African troops seized the area, business continued to flourish. The senior Erni, a Swiss national, was permitted to remain in business. The German soldiers were moved out to a prisoner-of-war camp; British and South African soldiers took their places in the hotel.

The fossil locality is marked by several rectangular stone structures, built by German colonial Schutztruppe for their horses (color plate 7). This isolated area protected the horses from a terrible scourge of rinderpest (hoof-and-mouth disease). In April 1897 rinderpest struck the cattle herds on the Schaf River south of Windhoek and spread like wildfire throughout the Namibian territory.[16] This particular spot has scant forage but it is near a spring.

The Schutztruppe, with their Australian outback-style hats, were fond of both horses and cattle. In Oskar Hintrager's 1955 book *Südwestafrika in der Deutschen Zeit* (*South West Africa in the German Era*) there is a playful photograph showing

Auf Ochsen berittener Zug der Schutztruppe 1904 [Schutztruppe squadron riding oxen-back, 1904]

Frau Seilacher pronounced *Schutztruppe* with affection, as if to say, "Those are our boys."

In the vicinity of the provincial Schutztruppe stables, *Pteridinium* and *Namalia* fossils are very abundant in an east-west striking quartzite bed associated with current lineation. Lines on the quartzite bed surfaces are suggestive of stiff currents. The *Pteridiniums* appear to be aligned; in one spot their alignment is in the same direction as in the current lineation in the bed above. In sediments nearby are rounded, flat mud chip horizons. Cannonball-sized antinodules weather out of the quartzite.

The sediments here are thought to have accumulated in the tail end of an alluvial plain and were deposited in shallow marine water.[17] The sandstones in particular originated as fan deltas and sandy fill sediment of tidal channels along a quiet water shoreline, perhaps indicative of the fact that the shoreline faced a narrow seaway of limited extent. Large sigmoidal foreset beds in the lower part of the fossiliferous Kliphoek Member are evidence of fan deltas. Gypsum crystals in sandstone and shales, mudcracks, shale clasts, and herringbone cross-stratification are all indicative both of salt water and tidal deposition.

We located a low ledge, underneath which were wonderful three-dimensional fossil specimens weathering out of a fairly crumbly facies of the sandstone. Seilacher was engrossed in this site and his considerable powers of three-dimensional reconstruction were evident. However, his abilities in this regard were strained to the limit because the *Pteridinium* specimens were contorted into a variety of fantastic shapes. My first specimen, 1 of 7/30/93, is a *Pteridinium* specimen that turns the corner (color plate 8); that is, it is folded back on itself. *Pteridinium* usually has two vanes parallel to bedding, with outer edges that curve up, and a medial third vane or "chaperone wall" that projects up vertically. The flatness is curved and deformed in some cases. Sometimes it looks as if the fossil is acting like an elastic sediment slingshot or sediment scoop. Specimens definitely can give the impression, as Seilacher maintains, that these things lived in the sediment or were subjected to violent burial. Trough cross-bedding is seen on some float blocks nearby.

On bedding sole surfaces (bottom of bed) are numerous *Pteridiniums,* with long axes in rough alignment.[18] In some cases the medial sutures curve strongly in parallel. Some specimens seem to step up from one bedding surface layer to the next (figure 4.3). I left the locality that day even more convinced of Ediacaran strangeness.

Once every 3 months a community dance is held in Aus, and we planned to attend this big event. Before going to the dance we met Clifton, a Bushman–white mix who was interested in improving his English. He introduced himself to me and sat down with us outside the bar and dining room (Eet Kamer in Afrikaans) adjacent to the Bahnhof Hotel. Clifton worked as a farm hand, making only 170 rand per month, and admired me and my group because I was, as he put it, "somebody." I told him Namibia is a beautiful country—"Aren't you proud to be a

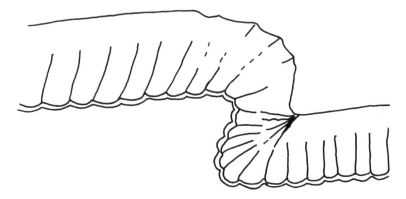

FIGURE 4.3: The chaperone wall of a *Pteridinium* specimen stepping up from a lower bedding plane surface to a higher. Height of chaperone wall 16 mm.

Namibian?"—at which point his eyes got teary and he replied, "Don't start with me on that!"

Seilacher thought that Bushmen have curious, compressed features with high cheekbones and a narrow lower jaw, a fetal face. He attributed the latter to neoteny. I replied that at some point bushmen women must have preferred such a face.

Because the upcoming dance is held only every 3 months, we expected a whole community affair. We soon learned that the dance and food (white bread-and-butter sandwiches,[19] pancakes rolled around a nut butter, and meat stew on rice with hot chutney) were for whites only. This was the first instance of discrimination any of us had encountered in Namibia. Whites seemed like Americans or Europeans physically, but I sensed a great cultural graben separating us from them. Handsome young men danced with an older lady, who was a cook for laborers. When the music began, Seilacher started up and danced with the hotel/bar manager, a rotund lady with a pleasant English accent. Frieder immediately took a dance with Edith Seilacher, and I danced with her myself later.

Frieder and I quietly nicknamed an older, ruddy-complected blond man in the crowd Socrates, on account of what we felt was his Grecian profile, where his downward directed nose approached his upward-directed chin. This man danced up a storm with the ladies, waltzing, jiving, anything. Frieder and I joked to ourselves, "I hope Socrates doesn't order the wrong drink!"

A half-breed Bushman stood outside the door to the dance hall with a hangdog look on his face. Later he came in to bus tables. The live music band consisted of a man on keyboard and another playing an intricate but slow tempo electric guitar lead. No vocals.

The morning of July 31 we returned to the Schutztruppe stable locality, intent on finding and excavating a large slab of quartzite bearing the *Pteridinium* fossils. We tried at first to excavate a low ledge where we had found the three-dimensional fossils yesterday, but we had no success; the more we excavated the low ledge, the less likely it seemed that we were going to find more fossils. So we left the exposed specimens intact and put back the large blocks we had pried out with the 167-cm pry bar. I spotted between the rocks a plant rootlet fuzzy with mycorrhizal fungi.

After a lunch of Windhoek bread, peanut butter, dry and moist sausage, pears, apples, and oranges, I trekked to the east of the locality and tried to follow the fossiliferous bed along strike. I saw some decimeter-sized flute cast-like bedding structures in the sandstone, and abundant rounded mud chips. Next I came upon an amphitheater-shaped depression in the bedrock, with a spring in the bottom of the depression. Sparrowlike birds were diving for cover whenever I moved, but they returned to sipping water and snatching flies when I sat down. I hiked past the amphitheater and thought about the prospect of being stalked by a leopard near this water source. Farmers in the area drive cheetahs and leopards into traps stretched across wind gaps. Cheetahs are bagged (literally) and sold alive. Leopards, more dangerous, are killed. Both species are endangered.

I headed back toward the rest of the field party and found several *Cloudina* specimens in a buff-colored limestone bed. The Schwartzrand Limestone is limy, as I confirmed with the hydrochloric acid fizz test. I came across a jumping spider and a giant cricket with harvestmanlike legs.

As I returned to my colleagues, Seilacher called out "Mark, we found it!" The party was preparing to excavate beds exposed near the head of a shallow wash approximately 150 m from the Schutztruppe stables, where the van was parked. At this new site, *Pteridiniums* were spread out on the rocks over a distance of about 10–15 m, and the blocks could be easily lifted. *Lithops erniana* lives in rock crevices nearby, with some specimens living in fairly deep crevices where direct sunlight lasts only an hour or so. By 4:00 P.M. a series of rectangular subslabs had been removed, and Hans began, with the help of all, to glue the slabs together with an amazingly strong, pastelike achemie glue. Casting would be difficult because the fossiliferous slab was broken up into multiple blocks that had to be aligned properly in order to make a faithful cast.

Occurring as float[20] fragments near these blocks are orboid "flying saucers" 1 cm in diameter. We could not find any in matrix but the float source is clearly local bedrock. Seilacher called them *Protolyellia*, an organism now added to the diversity list of *Pteridinium* and *Namalia*, a simple discoidal form composed of sand.[21]

During the driving this day I learned how Seilacher's older brother was a radio operator in a Nazi bomber aircraft Heinkel HE-111. He crashed the plane twice. He was interested in becoming a physician, but died on the Russian front. Dolf never flew in the war but wanted to fly with his brother. Dolf was not excited by the HE-111 but he would have liked to dive in a Stuka dive bomber.

Dolf also told the story about almost being bitten by a horned viper in southeast Egypt. While peering into crevices trying to see the founder colony of an extensive fossil oyster bed, he spotted the coiled snake, poised to strike and only inches away from his hand. Had he been bitten, there would have been about enough time to smoke one cigarette.

On August 1 we again visited Hellmut Erni and were joined by Dolf and Edith's friends Manfred and Catherine. Erni directed us to a small shed adjacent to the dusty Karakul sheep quarters. Inside this shed is a wonder of Ediacaran treasures, Erni's storehouse of fossils, a mecca for paleontologists interested in Ediacarans. Included in the collection are such things as large scorpions preserved in jars of alcohol. Seilacher and I took photographs of Erni's *Pteridiniums, Namalias, Rangeas,* and regular and "elephant's foot" forms of the *Erniettas*. Shepherds on foot, we are told, were first to find the *Rangea* specimens.

Later in the day we reached the Aar site. At 12:40 Seilacher began to give the grand tour; having already heard the presentation, I set off in a general southerly direction to study the stratigraphy of the Schwartzrand Limestone. The Schwartzrand Kalk (Black Rim Limestone) is a prominent unit in the Nama Group and, as its name indicates, forms a dark colored rim on the low hills and canyon walls of the region. The Aar plateau, which we explored later, is capped by an expanse of dark bluish gray limestone. From my field notes for August 1 is an informally paced (my pace is 0.9 meters per step) stratigraphic section:

DIRECT LINE	
TO STONE	
MONUMENT	~30°W, 5° dip
61 [PACES]	Schwartzrand Black Rim: dark gray limestone intraclast conglomerate

17	Buff ledge w/? *Epiphyton*
12	Thin buff ledge
26	Buff-red-blackish, striped buff and dark gray
33	Buff gray w/blackish desert varnish in layers
53	Gray to buff with iron-rich swirls, nodular w/ *Epiphyton*
18	Road
37	Thin yellow buff unit w/scattered iron nodules
66	Calcareous sandstone w/rounded mud chip horizons; thin-bedded to massive sandstone to quartzite, friable in places with *Pteridinium* and *Protolyellia;* casting site

This day we made a trip to the central portion of the Aar Plateau to view petroglyphs on the top of the plateau. We rode out to the site in the back of Hellmut Erni's truck, a Nissan Hi Rider 1 Tonner 2500 diesel. Erni had a box of Courtleigh Satin Leaf cigarettes on the dashboard.

Bushmen and their ancestors formed the petroglyph art by impacting the smooth surfaces of the plateau-capping Schwartzrand Limestone with pointed rocks. Each hit made a white dot on the dark gray limestone, and images were formed by a series of closely spaced dots. Judging from the weathering on the outcrop, some of the images must date back thousands of years. The quality of some of the art is outstanding, comparable to Lascaux or Altamira cave art. Not all of the images are ancient; one, with much less weathered white impact dots, bears the date 9/9/1927. Thus, this appears to be the oldest continuing art series.

Bold images of rhinoceros, giraffe, eland antelope, oryx, snake, mountain zebra, and elephant are scattered over the plateau surface. Some of the animal images have somewhat distorted proportions. Many of these creatures no longer live in this region. One image looks like a tunafish or a penguin swimming at full speed. Human figures are less common, but one shows a female figure with large breasts lying on her back. A male figure, with an erect penis, is running away from her. Does this depict an ancient rape scene? A petroglyph of a male authority figure (the chief?) extends his right arm forward in an apparent pose of benediction. Under his outstretched hand is a petroglyph of a wheel with eight spokes.

At an overhanging ledge nearby we viewed delicate paintings in red pigment of human figures. The earliest examples of this type of Namibian rock art date back 19,000–26,000 years. The geometric patterns are believed to have been inspired by hallucinatory experiences. Much of what we know about Namibian rock art comes from a group of !Xam San men imprisoned in Cape Town for livestock theft, murder, and other crimes.[22] German philologist Wilhelm H. I. Bleek acquired custody of the men. The men worked as Bleek's domestic servants, living in huts in his garden. They supplied Bleek and his sister-in-law, Lucy C. Lloyd, with tales of their !Xam tradition.

Bleek focused on the men's language, whereas Lloyd transcribed approximately 10,000 pages of !Xam folklore and myth.[23] Some of the rock art images, originally interpreted as scenes of the hunt, are now, in the light of Bleek and Lloyd's work, viewed as portraying a rainmaking ritual. In this important ritual, the !Xam saw the rain cloud as a giant lobopodlike animal, striding across the parched land with billowy legs of streaming rain. The task of the rainmaker was to entice this giant creature from its waterhole lair, lead it to high ground, and then slaughter it. The falling rain represented the blood shed by the slain rain creature. The rain creatures depicted in the rock art are always large herbivores (hippopotamus, antelope) but usually show strange proportions and features.[24] So in a sense, the original interpretation of the scenes as the hunt is correct, with the quarry being life-giving rain.

The Schwartzrand Limestone here is packed with *Cloudina* fossils, most of the shells here having been moved and broken by currents and redeposited as thick coquinas (lithified shell hash). A large boulder of limestone sits atop the plateau. When struck it rings like a deep-throated brass bell.

Dropping off the plateau into a nearby desert wash, punctuated by deep pools favored by game, we found mistletoe growing up and out of the trunk of an acacia tree. The tissue of the mistletoe permeates the entire acacia. Erni related how many acacia trees on his land have been killed by mistletoe.

The next day (August 2) I continued my exploration of the Schwartzrand Limestone as work at the casting site continued. I continued hiking south of the stone marker of the day before. In a piece of Schwartzrand float (1 of 8/2/93) within 200 m of the casting site, I found probable cloudinids in living position. I hiked to the Schwartzrand Cliffs to the northeast of the casting site. Here I came across evidence of ancient human habitation.

Our hominid lineage has a long history in this part of the world. In 1992 a group including geologist John Van Couvering found a 13-million-year-old jaw from *Otavipithecus,* the ancestor common to apes and humans, in Namibian strata. The fossil is the first evidence of such an ancestor south of the Equator. The lower jaw is about 3 in. long, and its teeth resemble those of humans and apes rather than those of monkeys. The oldest known fossil specimens belonging to *Homo sapiens* occur in caves in South Africa in deposits only 120,000 years old.[25] So it was with great interest that I examined these cliff shelter dwellings.

The shelters were formed by a semicircle of stones built against the Schwartzrand cliff. The shelters faced west, perhaps to give the inhabitants light later in the evening. They were safe from predators here and had the natural amphitheater with a spring nearby. Like tract housing, the shelters were spaced about 100 m apart along the cliff face. White quartz and chert flakes in the center of the stone semicircles indicated that the dwelling builders were a Paleolithic people. At one point the cliff was overhanging, providing a particularly nice shelter. Ostrich egg fragments were scattered amid the knapped white quartz flakes.

That night, Bruce Runnegar (his name is Old Norse for "rye field") of the University of California at Los Angeles and Jim Gehling of the University of South Australia arrived in a white Volkswagen minivan nearly identical to ours. The next morning (August 3) we set out to the Aar locality to do the casting.

En route, as in earlier days, we encountered a family of ostriches. Mother, father, and six youngsters watched us warily from a distance. On average, five of the six juveniles would be taken by jackals before they reached maturity. Every day we saw the ostrich family in the same place.

Large birds have a long history in Namibia. A joint Franco-Namibian expedition announced in the December 1995 issue of *La Recherche* the discovery in the Namib desert of a 17-million-year-old giant egg. The egg was discovered by Brigitte Senut of the National Museum of Natural History in Paris. She found the egg, nearly intact, partly embedded in sandstone, and called it "one of the miracles of fossilization."

The egg had a volume of 1.7 L, as compared with the 1.2-L volume of the average ostrich egg, and a shell twice as thick as ostrich eggshell. The early Miocene egg, along with bones of early rodents, antelope, and an elephant-like proboscidian, were discovered in the Sperrgebiet ("forbidden zone"), until recently the exclusive preserve of diamond miners. Senut and her colleagues considered the possibility that the egg belonged to a large turtle, but the calcitic composition of the shell links it to the shells of other birds.

The egg has been attributed to "Namibia's ancient big bird."[26] However, no one knows how big this bird was or what it looked like. The species was nevertheless given the name *Diamantornis wardi* (Ward's diamond bird), to honor South African geologist John Ward. Ward studied ancient dunes of the region.

With Runnegar and Gehling, colleagues from the Namibian Geological Survey, and C. K. "Bob" Brain of the Transvaal Museum in our company, we were in high spirits driving to the field site. I asked Jim Gehling, "What would happen if a holistic thinker were to meet up with a reductionist?" Jim replied, "Annihilation!" Seilacher was in one of those moods that induces him to start telling self-effacing Swabian jokes. The Swabians are a tribe from the Swabian region of Germany to whom Dolf owes allegiance.

A Swabian mountain climber in the Alps is beset by a tremendous avalanche. After hours of frantic effort the rescue team finally digs him out. On coming to and seeing the uniforms of the Swiss Red Cross, the Swabian declares, "I gave at the office!"

At the site I took Jim and Bruce to the stone ring shelters. We examined sedimentary structures of the Schwartzrand, and noted thrombolites[27] in the buff-colored carbonate rocks below the Schwartzrand. A desert roach crawled across a rock.

Seilacher began with an exposition of his hypothesis of *Pteridinium* reproduction. He noted that there are forms living in different levels in the sediment, that the ones in the higher levels are the progeny, and that the edge of the side wall of the older generation "coincides exactly with the median line of the next generation." Runnegar asked whether the geometric arrangement could be mere coincidence and whether multiple examples of this relationship were known. Seilacher replied that yes, there were several examples, and that in this place the specimens were oriented in the same direction as the ancient water currents of the locale. Gehling asked whether this inference was checked with measurements of cross-bedding (a way to determine ancient current direction) in strata higher up in the stratigraphic section. I replied yes, they had been checked.

But because the fossils under immediate scrutiny were, under Seilacher's interpretation, buried in life, they would not have felt the current directly. Seilacher infers that the orientation of the *Pteridiniums* was in response to currents in pore water induced by water flow above the sediment surface. Furthermore, the next generation, growing as a bud along the edge of the outside wall of the parent, would develop in parallel to the orientation of the parent generation. Gehling then noted that some of the progeny *Pteridiniums* were oriented at 90 degrees to the adults. Seilacher replied that the grooves of the *Pteridinium* (at right angles to the axis) might also be able to align with current. In any case, in his view the orientation of the *Pteridiniums* was not caused by mechanical current transport and deposition of the bodies. Gehling seemed skeptical about this idea.

Runnegar noted that he was trying to make similar inferences regarding a slab bearing Australian *Phyllozoon* fossils, currently under assembly for transport to a museum. I replied that these *Pteridinium* slab specimens were to remain in Namibia, and that we were going to return with only a silicon mold of the slab specimens. Seilacher added that the actual specimens would remain on Erni's farm, and that Erni would be mighty proud of them.

Runnegar was asked whether he had a compass orientation on the Australian *Phyllozoon* slab. He replied that he did not. The fragments of the slab were transported by helicopter and assembled in his laboratory. Seilacher replied that, without the advantage of helicopter time in Namibia, his team was able to get an orientation on the *Pteridinium* slab. At this comment everyone broke out in laughter.

Seilacher later described the morphology of *Pteridinium* as a bathtub for unmarried couples, consisting of two troughlike bathtubs on either side, with a "chaperone wall" between, effectively separating the "couple."

Later I pointed out a *Pteridinium* specimen with an interesting profile. One bathtub wall rose steeply, and the other flattened out. Seilacher replied that this was commonly seen, with the bathtub wall flattening out like a ray's wing. I agreed that it was indeed much like a ray's wing, complete with the upturned fold right at the edge. Seilacher noted that he takes this kind of information very seriously because it provides three-dimensional detail.

Hans Luginsland skillfully guided the hands of the field party as they nestled the glued-together slab into a bed of sand near where the blocks were quarried. Hans mixed up the batch of silicon, combining the whipped-cream white silicon body with the navy blue liquid catalyst. The mixture smells like spackling. He carefully applied it to the base of the inverted sandstone bed, working the sky-blue material into every fold and flute of the fossils and every crevice of the rock, ensuring that no air bubbles had formed. It was very much like frosting a cake, and Hans used a pastry brush. This led to a discussion of German bakers and German bread. Seilacher noted that authentic German bread is available in New Haven, Connecticut, but must be flown in from Canada and costs $7–8 a pound. When Hans's work was completed, all that remained was to wait for the curing of the silicon mold.

We had had rain recently, and unfortunately that night proved to be unusually cold. Although this was the desert in a generally warm region, it was nevertheless the austral winter. The silicon was not designed to cure at freezing temperatures, and the cold not only stopped the curing process but ruined the silicon so that it would not cure at all. This disastrous state of affairs was not discovered until the next morning.

On August 4 Hans was in a state of high agitation. His first attempt at casting, the main object of the expedition, had been ruined by an unexpected frost. What to do next was not clear. Jim Gehling recounted how he had attempted to make a mold of an important Ediacaran fossil in Australia, one still attached to bedrock. Something had gone wrong with the molding medium, which turned to an inflexible, immovable glue and remains attached to this key specimen to this day.

Much discussion was spent on what to do with the *Pteridinium* slab, covered with slimy, ruined silicon. Finally it was agreed to build a wind shelter to help keep the rock warm, scrape off the old layer (a messy and laborious process), and apply the next layer with a more than ample charge of catalyst. Fortunately, the Seilacher team had brought along extra silicon for just such a contingency. The new silicon layer was applied where the old had been, and we all hoped for the best.

The day of the silicon problem I traveled with Gehling, Runnegar, Friedrich "Frieder" Pflüger (Seilacher's graduate student), and Brain. Our object for the day was to relocate the type locality of *Ernietta plateauensis*. We pulled off of a paved road near a farm windmill and what appeared from a distance to be an anticline. As we hiked in to the locality, we were excited to find a black chert. Black cherts sometimes harbor exquisitely preserved microfossils, but later work showed this layer to be unpromising because it had recrystallized, obliterating any fossils that might have been present.[28]

As we approached the *Ernietta plateauensis*–type locality on foot, Pflug's point C in his 1966 paper,[29] we encountered a troop of baboons. The alpha males challenged us from across the canyon, and we traded calls with them for several minutes. It was the first time I had ever attempted to communicate with another primate species in the wild. Apparently they got the message, for they left us alone for the rest of the day.

While hiking we regaled one another with earthy field stories and were subjected to Bruce Runnegar's particularly acrid sense of humor. Someone suggested expanding Namibia's culinary spectrum by opening a specialty shop for sheep-dung-maggot shish kebabs.

We hiked into a small canyon and over to find a fault contact between the quartzite and limestone. Soon we reached the ledge marking the contact between the Daris and the carbonate rocks, and realized that this must be Pflug's point C. Brain took a photo (figure 4.4) of three of us (me, Bruce, and Jim) on the site. We found no erniettids, but I did find two *Pteridinium* specimens. The first (2 of 8/4/94) was a nice three-dimensional piece with vertical chaperone wall intact. The second was poorly preserved but showed paired bathtub walls. Both were preserved in sandstone. This is apparently a new *Pteridinium* locality.

Bruce Runnegar tried to talk me out of the first specimen, for he correctly realized it to be a specimen of potential importance. I politely refused, saying that perhaps we could talk about it later.

Next we drove to the Kuibis area. There we met a Mr. Loots and asked him for permission to go to the *Rangea schneiderhoehni* type locality. He told us that the land was owned by a Mr. Blow and was called the Aukam property. We located a narrow quartzite ridge near a railroad track and the edge of the Loots property. We hiked up to the trig station (tall aerial and solar panels) and over the hill but found no fossils. The glaring white quartzite crops out only irregularly here; the rest is a jumble of rounded boulders. We headed back and stopped just past the north end of the ridge. Getting out of the van, I found a poorly pre-

FIGURE 4.4: From left to right, the author, B. Runnegar, and J. G. Gehling in the field in Namibia in August 1993. The Schwartzrand (Black Rim) Limestone is visible in the background.

Photograph courtesy of C. K. Brain.

served *Pteridinium.* Runnegar was particularly pleased with this; at least we could confirm that fossils occur at this site. Now we knew that *Pteridinium* co-occurs at the type localities of both *Ernietta plateauensis* and *Rangea schneiderhoehni.*

We returned to the Bahnhof Hotel, and I was feeling quite pleased with myself on account of the day's discoveries. The small living room of the hotel had a fireplace, and as we gathered around for evening drinks we began the Ediacaran debate. This had been planned by Seilacher to be largely a debate over the validity of his Vendobionta theory, with Runnegar playing the role of devil's advocate. Seilacher had it set to be a contest between the reductionist and the holistic points of view.

Seilacher began by asking whether we all agreed that life of the Vendian is a phenomenon all its own—not merely an extension of the Cambrian world, but a phenomenon unto its own self, with a unique character throughout the world. Runnegar affirmed that all present were in agreement with that view, but then he refined the question by asking whether the Ediacarans were monophyletic, that is, members of the same group of related organisms. Seilacher agreed that, if so, this would make the Ediacarans even more unique. Runnegar then asked whether the organisms were similar because they were responding to similar environmental circumstances of the time rather than all being closely related. Seilacher replied that it was necessary to make exceptions to the Vendobionta scheme right away. Nevertheless, it was his preference, as far as possible, to treat them as all part of the same group. This would be in contrast to other paleontologists who would consider each form separately, suggesting that *Dickinsonia* resembles a fungiid coral and ignoring the rest of the Ediacarans.[30]

Seilacher's main exceptions would involve Vendian organisms that do not fit his Vendobionta model. For example, in Newfoundland there are specimens of unequivocal Vendobionts (such as frond fossils), but there are also forms variously and informally called lion's feet, dog's feet, and so forth. These forms are merely roundish globs, or globs within a glob, so under the traditional phylogenetic scheme they become assigned to the jellyfish group. Thus, the traditionalists have a ready explanation for everything.

Seilacher continued by noting that the glob forms do have a morphology, but it is a morphology of "dumplings in a plastic bag." He would not include such things among the quilted Vendobionts. Similarly, the large pogonophoran-like tube worms fossils from Ediacara, Australia,[31] would not be part of Vendobionta.

Runnegar asked whether the sand corals and trace fossils would also have to be excluded. Seilacher replied that those are different, that body fossils were currently under discussion, and that sand corals were in a different class altogether. Seilacher then reminded Runnegar that the Vendobionta concept includes not only the unique body construction of the Ediacarans but also their unique preservation. Runnegar objected that he would not wish to include *Tribrachidium* and

Spriggina along with the other Ediacaran body fossils. He continued by saying that it might be reasonable to test a hypothesized family relationship between, say, *Phyllozoon* and *Dickinsonia* (which have enough similarities to make such a comparison possible), but that it would be problematic to test phylogenetic similarities between all of the Ediacaran forms.

Not so, replied Seilacher, for years ago he had hypothesized that *Spriggina* is merely a variant of *Charniodiscus*. Runnegar responded that smaller taxa such as *Spriggina* had a very different type of preservation. Seilacher replied that, on the contrary, he had seen these smaller taxa on a slab with *Dickinsonia*, sharing exactly the same kind of preservation. Runnegar acknowledged that they are indeed on the same slabs. Seilacher continued by noting that he saw in the Ediacarans a sequence of budding, growth that is bipolar or unipolar, and no legs or any other type of organs, and furthermore no differentiation. Runnegar claimed that the discussion was not going anywhere because of differences in interpretation of the same fossils, to which Seilacher agreed. Runnegar added that in his opinion, *Spriggina* and especially *Tribrachidium* were very far removed from Seilacher's concept of the Vendobiont air mattress style of construction. Seilacher replied that in his opinion, these forms could be reconciled with a Vendobiont placement. In *Tribrachidium*, he sees two orders of element bifurcation, the most distal of which could be a type of quilting. The coarser (earlier) order of bifurcation looks to Seilacher to be very much like the stem sections of *Charnia* or *Charniodiscus* in Newfoundland. He noted that this may indicate a different kind of material in these parts of the bodies of *Tribrachidium*, *Charnia*, and *Charniodiscus*, perhaps indicating the presence of a more solid or gel-like consistency, in contrast to the more biologically active, foliate parts of these organisms. I asked Seilacher whether he was suggesting that *Tribrachidium* represented a fossil holdfast. No, he viewed it as a complete organism with three strengthening radii forming the basis for quilted, foliate parts of the creature. Gehling added that he was finding that the Ediacarans had another level of structure, overprinted on the primary structure that Seilacher had just described, including "strange fanlike structures radiating out over" the branches and subdivisions of the branches themselves. Sometimes these finer features were not preserved at all.

Seilacher acknowledged that there were disagreements, then proposed that we provisionally call the organisms Vendobionta whether or not we all accepted the phylogenetic implications of the term. Runnegar added that he was willing to use the term *Vendobionta* without provision but would use it only for four of the genera. Runnegar objected to Seilacher's shoehorning of most of the other Ediacaran taxa into the Vendobionta. Seilacher agreed to disagree.

On August 5 we drove to the casting site.[32] Silicon had not yet set because of a frost the night before. I took Dolf and Edith up to the two archaeological sites. Then I drove with Erni to the Pflug locality between the casting site and yesterday's first site, the type locality of *Ernietta*.

Once again, the fossil horizon is at the top of the quartzites just before they give way to buff carbonates. We saw Flädle structures[33] similar to those seen just before the first fossil find of *Pteridinium*. I caught, mesmerized, and released an *Agama* lizard (family Agamidae).[34] Back at the casting site we found clusters of *Protolyellia*. Jim Gehling found a *Paramedusium africanum* Gürich 1930 (figure 4.5) in fine clastics just downsection from the main *Pteridinium* bed. This, according to Jim, was a happy find because the type specimen of *Paramedusium africanum* was lost during World War II.[35] I found a strange specimen that we dubbed the wrinkled frond.

According to measurements by Jim Gehling and Frieder Pflüger, the axes of the *Pteridiniums* trend north-south; paleocurrent indicators are to the southwest.

Later in the day Seilacher entertained us with stories of his past exploits. Off the coast of Sudan, in a scene reminiscent of Captain Nemo's dive in *Twenty Thousand Leagues Beneath the Sea,* Seilacher cut himself on the edge of a giant *Tridacna* clam. Another time, astonished tourists in a glass-bottomed boat peered through the glass to see Seilacher busily at work, diving on the seafloor.

FIGURE 4.5: *Paramedusium africanum* from the *Pteridinium* casting site in Namibia. Ediacaran medusoids such as this one are rare in Namibia. This specimen was collected by J. G. Gehling. Scale bar in centimeters.

The night of August 5 was very cold in the hotel; no heat. We felt the bite of one-star accommodations. Before retiring we had a lot of red wine, conversation, and Jim Gehling's splendid photographs of fossils from Australia.

The morning of August 6 we planned to drive to Lüderitz on the coast. Peter and Hans checked the tire pressure in the Volkswagen minibus at the Aus Namib Garage (Souvenir, Koeldrank, Sigarette). A Trans Namib oil truck pulled up beside us. A poor black child with holes in shoes looked on. He looked cold. I gave him some cash.

It is 125 km from Aus to Lüderitz. We passed rounded granitic outcrops west of Aus. My attention was captured by a spheroidally weathered granite dome monolith. Edith Seilacher commented that it must be a large ostrich egg. The terrane of this area is not unlike the rounded granites of Joshua Tree National Monument in California; another boulder-hopper's paradise. We spotted two ostriches on the side of the road, under the "egg."

Next dark conical hills appeared, protruding from a tan plain dotted with trees. Bush turkeys glided to the left, seeking cover in the hexenhazel (witch hazel). A dark trapezoidal massif rose to the right.

A cautionary sign was posted beside the road:

Warning. You are now entering the sperrgebit [sic] diamond area No. 1. You must not move to left or right of road without permit by order of the Diamond Resources protection statute.

Are the dark hills kimberlites, I wondered? A jackal to the left of the road looked a great deal like a North American coyote. The hills to the left of the diamond area are marbled with browns and pinkish tans of the Namaqua Metamorphic Complex. Sign on right:

Namib Feral Horse

The Schutztruppe's horses were released after they were captured, and went feral just like the mustangs and feral burros of the American West.[36]

The dark trapezoidal mass continued to loom off to the right. The Gorub station appeared at the left. It is the first station after Aus. Railroad construction engineers had a habit of putting stations at regular intervals, every 15–20 km, to service steam engines.

A single springbok was seen to the left. The vegetation all but vanished as we crossed a sandy, pebbly plain. There were some tire tracks and stubble beside the road, but that was all. Telephone lines ran parallel to the railroad tracks on the left. Power lines intersected and marched southwest through a wind gap that looks like Dr. Seuss's West Jehosephat in *Oh the Places You'll Go!*[87]

Lüderitz was now 80 km distant and grass had returned to the plain. Sand dune crests were visible to the right of the road; they merged smoothly into the pediments that support the distant mountain ranges. We stopped for a photo. Frau Seilacher pointed a Blaupunkt video camera. Peter, who had been driving, aimed his Pentax. Being low on film, I retired from the vehicle simply to relieve my bladder. My friends cried out: "Diamond area—Not allowed! Not allowed!"

Tasteless humor seems endemic to geological field work. Herr Seilacher told of a trip to Jordan; as a participant turned away from camp to pray to Mecca, an American followed him, unwittingly, to urinate. Dolf had to stop the American.

Seilacher continued: A well-known Swiss professor was much admired by students, who followed him around, making remarks to try to impress him. Finally, they followed him into a crevice in a canyon that got narrower and narrower, until he finally exclaimed, "You don't have to follow me for this!"

An orange and green Leyland truck passed us on the right, going in the opposite direction. The Tsaukuib Station appeared on the left, the name derived from the Bushman language. A blue and white bus pulled off of the road ahead; as we passed we saw that it was a Safaris Limited bus filled with South African tourists. More possible kimberlites to the right—black hills with low relief—surrounded by a black and tan alluvial fan. One hill, apparently not a caldera, nevertheless looks like a Namibian Diamond Head.

The dunes were more visible now and had an orange color with a band of blue ocean beyond. The vegetation became sparse again. The kimberlites in this area are not the appropriate age for diamonds; most diamonds are transported down the Orange River mouth. The diamonds are carried by currents along the coast. There are lesser quantities of them to the north of the mouth, but the highest-quality diamonds are found in the Namibian north coast because of a natural sorting process.

A game ranger who had caught a baby ostrich and carried it to the hotel in Aus in a cardboard box a few days ago (it was quite cute next to the fireplace) remarked to us that ostriches have been killed by hunters seeking diamonds in their gizzards. The diamond gastroliths, being of course harder than all other rocks, preferentially survive the gizzard grinding phase of the digestive process. Some very large and valuable diamonds have been collected at the expense of ostriches' lives.

Written in white rock on hills to the left is the cryptic message

KR
NUNG ANIMUS

A few *Aloe dichotoma* were seen scattered on hills across the road.

We saw more hills reminiscent of Joshua Tree National Monument. Could a variant of the Joshua tree principle apply to landscape recognition? Do similar landscapes evoke uncannily similar vegetation? The South Atlantic was beginning, as we proceeded, to fill the low spots in the western horizon like a rising tide. We passed a battered and stripped small blue station wagon to the right. Sand dunes were beginning to drape the hills to the left. Termite mounds and grass tufts were scattered on the treeless plain.

Lüderitz was 30 kilometers away. Last time Frieder was in Lüderitz, the harbor held part of the Portuguese fishing fleet, captured for fishing within the 20-km exclusive fishing zone of the new Namibia. The boats were still for sale.

We passed the Rotkop Station, marked by a sign but no structures. A transformer station appeared to the left. The dunes rise high to the right-cuspidate dune forms open to the southwest. A utility truck was off to the right. A sign cautions drivers about wind and sand:

100 km/hr
60 w/sand

Gray sand was indeed streaming across the highway to the right. Another sign:

Private C. D. M.
Lüderitz 20

The founder of the coastal town, Adolf Lüderitz, was a merchant adventurer born in Bremen to an eventful life. He spent a few years in the United States and returned to Germany intent on colonial expansion. He applied in 1882 to the German government for protection of any acquisitions he might make on the Namibian coast. The government in Berlin quietly approved his plans, and in 1883 Lüderitz set sail in the *Tilly* to Angra Pequena.[38] After purchasing land from the Namas and the Hottentots of Walvis Bay, he established "Lüderitzland." In the opinion of *Reichskanzler* Bismarck, it was time for Germany to stake out a "place in the sun" and establish the first German overseas colony. In 1885 Lüderitz founded the Deutsche Kolonialgesellschaft für Südwestafrika (German South West Africa Colonial Company).

The company was founded with funds raised from German investors as the company went public. Lüderitz himself purchased all the land and mining rights of Lüderitzland. He had incurred huge debts by loading up the *Tilly* to found the colony. All of his profits were reinvested; Lüderitz himself never became rich. He was drowned in 1885 in a sailing accident between Angra Pequena and the Orange River mouth.

The diamond fields were discovered in 1908, the same year that the first diverse Ediacaran fossils were found in German South West Africa by P. Range and H. Schneiderhöhn. The first diamond deposit was found by Zacharia Lewala, a railway worker, while shoveling drift sand off of the line south of Lüderitz. A diamond rush by white settlers was checked, however, by State Secretary for the Colonies Herr Dernburg, who was in the country at the time of the discoveries. Dernburg placed the diamond deposits in the hands of a company appointed by the German government. The Sperrgebiet, a 100-km-wide coastal strip, was the exclusive domain of the German South West Africa Colonial Company, with the diamonds being marketed by the Diamantenregie des Südwestafrikanischen Schutzgebiets of Berlin. Diamond production went from 38,000 carats in 1908 to a million and a half carats in 1913. Production from 1908 to 1913 was 5 million carats, which contributed 60 million marks to the German treasury.

There were bitter feelings toward Dernburg from the colonists, who were deprived of the chance to find diamonds on their own. Nevertheless, the finds made the German colony solvent for the first time and allowed Germany to spend funds for welcome improvement of the colony's infrastructure. The rallying cry among the more sensible farmer colonists was, "We must turn our diamonds into water."

We passed the Grasplatz Station, a tan building with a red roof on a bedrock and stone pedestal. Doors and windows were gone. The place looked deserted and was surrounded by sand and rock. *Grasplatz* means "lawn." It reminded me somewhat of the Norse naming of Greenland.

Sand streams were making it all the way across the road, and the sea stretched all the way across the western horizon. The Joshua tree granite landscape was drowning in sand. Their wooden ties not visible, the railroad tracks emerged from the sand like paired, parallel iron serpents. Spindly skeletal trees rose from the sand sea. Barchan dunes appeared to the left, and we saw a willowlike plant with leaves of leather. Examining a tan seed pod, I found two bugs, each marked with a black and red *X*. The Barchan dunes, opening to the west, tried to march across the road as we continued on. Local vegetation looked like tumbleweeds that hadn't learned yet how to roll. Lüderitz 10 km. A small airport appeared to the right. We passed the remains of Kolmanskop. A major casino formerly run by Erni's grandmother, it was a ghost town. Kolmanskop's attractive colonial buildings were missing windows and roofs.

Cresting a ridge, we caught sight of Lüderitz. The harbor looked like a lake because its connection to the open Atlantic could not be seen. Ancient Phoenicians, taking orders from Pharaoh Necho in 600 B.C., are said to have circumnavigated Africa in 3 years.[39] If so, they were probably the first Europeans to see (or at least sail past) what is now Lüderitz Bay. The great historian Herodotus dutifully reported, but doubted, the Phoenician report to Necho that halfway through their voyage, the noonday shadows pointed south.

Lüderitz looks like a frontier mining town with a fresh coat of paint. A white water tank is visible to the south. Mokolian biotite-rich banded gneisses, again of the Namaqua Metamorphic complex, form the outcroppings of the rugged local landscape. A yellow garbage can proclaims "Diamond Area Keep Out." Edith Seilacher calls this place a moonscape with electricity (color plate 9). The Phoenicians would have sailed past in all possible haste. The Lüderitz golf club appeared on the right. Edith, her humor in rare form, noted that they must have wicked sand traps.

The Namibian flag flies over palms and trees and corrugated tin roofs. The "harbor" is actually a lagoon. We saw signs for Bismarckstrasse, Saddle Hill Namibia Fishing Company, and Dial a Movie. We park on Diazstrasse. Blacks and whites are talking on the street.

We entered a small grocery store (Bäckerei Celbrodt). A banner read "100 Jahre Lüderitzbucht 1883–1993" [100 years Lüderitz Bay 1883–1993]. A colonial style trim of carved wooden leaves ornamented the edges of the walls. A picture of the founder was on the wall, with a turned-up mustache. The proprietor was black, spoke German, and sported sideburns and a baker's cap. I bought a package of SAD (South Africa Dried Fruit Co-op, Ltd.) Safari Pitted Dates, Produce of Iraq.

The Seilachers call the colonial building style in town Wilhelminian. This neoromantic style was elevated to the Empire Style through the direct influence of Emperor Wilhelm II. It is best expressed in the creations of architect Franz Schwechten (1841–1924). It looked to me like an attractive merger of Dutch and Tudor style.

We slipped into the First National Bank to change $1000 U.S. to rand for Bruce Runnegar. We received R3280.86 for the $1000, with commission of R33.14. Interest rates (*Rentekoerse*) posted on the wall were well over 12 percent, evidence of a capital-hungry, high-risk economy. Bank money exchange occurred in a office/booth with darkened glass. The office had bare tubular fluorescent lighting and a Westpoint air conditioner wall-mount. Also wall-mounted, on a shieldlike wood plaque, was a taxidermic preparation of a spiny lobster. It looked much like the species native to California waters. Pen and ink sketches of hoofed animals hung below a green sign showing a man running, presumably an exit sign, but it seemed to point to the office of the Manager (*Bestuurder*).

Fliers at the bank: "Here's what you should know about South Africa's new R50 banknote," "South African Reserve Bank—Money you can be proud of." "New banknote [R50, lion; R20, stately elephant] incorporates many highly advanced security features, making it extremely difficult to forge." Back of the new R20 bill depicted mining with a diamond intaglio. The R50 bill showed carbon atoms bonded into the covalent diamond structure. Leaving the bank, we passed Diamontbergstrasse.

We hiked up to the famous Lüderitz Gothic Church. It sits on honeycomb-weathered granites and wildly folded gneisses and mafic dikes of the Namaqua Metamorphic complex. The migmatite-gneiss swirls and the quartz-feldspar pegmatites are discretely stabilized by concrete. A jet black skink crept out of a joint between the concrete and the rock. Pink and white flowers were in bloom at the top of the outcrop. The rock surface glistened like a surf-washed gem.

From here we could see 20 fishing boats rocking in the harbor, all with bows pointed inland. Two larger trawlers were visible closer to the mouth of the Bay. A boom on the dock serviced the ships. Offshore ran the cold water Benguela Current; it reaches only 15°C during the hottest part of the austral summer. The seawater was blue-green with a brown tint, identical in color to the cool water of the Kuroshio/California Current offshore coastal California. It even had kelp.

The Baja California–like landscape had very sparse vegetation. Broken bottles littered the base of the honeycombed granite outcrop. The gables of the Wilhelminian-style homes and public buildings were faceted like diamonds. The buildings were painted in the tans and "southwestern" pastels that are so fashionable in southern California. One modern building seemed to combine Spanish and German colonial styles. Red roofs came in stucco, shingle, and corrugated sheet metal.

We continued on foot through Lüderitz. Krakenhaft Lampe. Boekwinkel on Nachtigal (=nightingale) Strasse. Livingstone Reiches Apotheke. Caltex oil tanks. New Institute for Fisheries for Department of Works. Monkey puzzle (southern hemisphere genus *Araucaria*) trees. Kapps Hotel. Black citizens were mostly of the Ovambo ethnic background.[40]

Lüderitz was an important city in the 1920s but it has been largely eclipsed by Walvis Bay. Although its influence is still apparent, the German language is dying out. Three flags are flying here: German, Union Jack, and Namibian. Lüderitz is quite isolated, as access is only by sea, small airport, or the narrow and sand-dune-encroached highway from Aus.

En route back to Aus, a pediment within mountain ranges to the north looked like an inland sea. Another aquatic mirage shimmered on the highway in the distance. Peter Seilacher drove at autobahn speeds—nearly 140 kph. The mid-afternoon light gave the usually straw-colored grasses a green sheen.

A landmark marker appeared in the distance, a white rectangle surmounted by a black square. Surely of use to travelers in the trackless reaches of the Sperrgebiet.

We pulled over to watch a gemsbok to the right. It galloped off, displaying a black-and-white rump and a streamerlike black tail. Seven more gemsbok appeared, resembling caribou from a distance. Their long, straight horns glistened, in the words of Henno Martin, like burnished swords.

At 20 km to Aus, trees were clustered on the low ridges like California coast live oak with spreading, fractal dendritic limbs. However, these were acacias with pastel blue blossoms. Large bush turkeys flapped across the road. Ten kilometers from Aus, on the right in the Joshua Tree Monument–like granite hills, small trees or large green bushes grew preferentially along the contact between the granite bedrock and its rubbly talus. Do the roots slope away from the firm bedrock toward a water source ponded at the base of the talus?

Passing into the Aus suburbs (Aussen Bitilk), we drove by the Aus shopping center, architecture in faux gothic. We were now in the flat-topped Nama Group mesas and en route to the Aar and Plateau Farms. A sign said "Gravel on Road," and indeed the road was entirely gravel. We were 100 kilometers from Goageb.

Areas around Lüderitz and Bogenfels were worked by several German diamond mining companies until 1920, when Sir Ernest Oppenheimer bought control and amalgamated these interests into a new company, Consolidated Diamond Mine (CDM) of South Africa. CDM, the largest single contributor to Namibian national income, is a subsidiary of De Beers Consolidated Mines, leader of the world diamond industry.

German mining interests sold out to CDM for 40 million reichsmarks. The money later became worthless because of the hyperinflation of the Weimar era. At the time of the sale, the diamond resources were thought to be running out. In 1925 diamonds were found south of the Orange River, near Port Nolloth.

A Dr. Hans Merensky established a link between the diamonds and fossil oyster beds in the nearshore area.[41] The oyster beds acted as a baffle trap for the diamonds being carried by currents north along the Namibian coast. The diamonds worked their way into the crevices between the oysters, turning the beds into a paleontological equivalent of Jason's Golden Fleece. In 1928 CDM discovered rich deposits in marine terraces just north of the Orange River, over 100 km south of the original German workings.

The diamonds are believed to have originated in volcanic pipes (kimberlites[42]) far in the interior. A swarm of kimberlites cuts through Proterozoic rocks of the Gariep Complex (mostly dolomites, shales, and their metamorphic equivalents), the sedimentary rocks underlying the Nama Group. In R. M. Miller's geological map of Namibia,[43] these kimberlites are 10 to 20 km due east of the coast, opposite Black Rock Island. The diamonds were carried to the sea by ancient rivers, then thrown back on the beaches by the Atlantic waves.

At 90 km to Goageb we saw ancient Precambrian igneous and metamorphic rocks of the Namaqua Metamorphic Complex frosted with the nearly flat-lying quartzites of the Daris Formation, in turn overlain by the carbonates of the Nama Group. Likely prospects for Ediacaran fossils could be seen at quite a distance because the fossils occur right at the break in slope caused by the bedrock transition from the quartz-rich rocks to the limestones.

Windmills were as common as trees. Inverted concrete *U*s were grouped in clusters, awaiting the macadam transformation of the road gravel. The Ovambo workers of the construction crew wore gray jumpsuits with ski caps. Komatsu graders were leveling the road surface, and red dust was everywhere. The Komatsu worked with a sand mover Caterpillar and a D9 Cat. Speed limit on the gravel was 60 kph.

We crossed a double *X* railroad crossing, which brought us to a sign announcing the Plateau Farm:

Plateau

H. Erni

Power lines crossed the road. The pad stretched ahead in dusty red. The depth of the dust made for tricky driving. I had skidded off the road slightly here earlier, and promptly turned the van over to Hans, a more skillful driver. The washboard went all the way across the road in places. Abundant grass seeds on the edge of the road looked like a dusting of snow. A blue water truck was spraying the construction area to keep the dust down. A horse stepped leisurely out of our way and off the road.

The road became increasingly rocky as we gained altitude. The dominant plant here is the spiky *Euphorbia,* with its poisonous milky sap. The road crested a rocky rise. A fence stretching across the plain caused a grazing discontinuity in the grassland. Soon we were back to the red silt road surface. Green and ochre vegetation was visible on quartzite slopes. Gray bushes, straw grass, and reddish termite mounds covered the grassy plain; bitter melons the size of softballs were seen in the road. One mound had its top smashed in; perhaps the site of a baboon snack? In the approach to Plateau Farm, a windmill turned slowly on an *Acacia/Opuntia* oasis. Peter, driving now, braked for birds and fishtailed the van in the silt. Frieder joked that 20 years from now, Nama children would say they were born *x* number of years after Mark McMenamin ran off the road.

Plateau Farm supported statuesque prickly pear (*Opuntia*) and trotting heifers. We arrived at the stone farmhouse, corrugate sheet metal roof gleaming, 41 minutes late for our rendezvous. Hellmut Erni's wife, very young, greeted us. Dolf, Bruce, Jim, and others were even later than we were. The farmhouse grounds formed an arboretum of pines, acacia, and spiny and smooth organ-pipe cacti. Hellmut's mother-in-law, with white hair and a brown sweater with a tan stripe, joined us on the porch. I played fetch (using a pine cone) with an energetic black-and-white sheepdog named Fips until a much larger but limping Rottweiler named Max tired of our antics. The Rottweiler chewed up the pine cone. Later Frieder played catch with both dogs, and the Rottweiler got quite winded. Fips was a sheepdog but was in fact more of a house dog.

Soon the others arrived and we paid our respects to the famous Erni collection of Nama fossils, kept in a shed by the Karakul sheep. The Karakul lambs were frisky and had climbed up on top of the corrugated metal roofs of the stone stables in order to nibble acacia leaves. This was our second visit, and the qualities of the fossils continued to amaze us. The end of one *Pteridinium* swelled downward like the rounded body of a Precambrian mandolin (color plate 10); a straight, thinner *Pteridinium* nearby could have passed for the imaginary instrument's fretted neck (color plate 11). Bulbous specimens of *Ernietta* defied our attempts to understand *Ernietta*'s mode of growth. The specimens appeared to be kinked, wrinkled, and swollen like a water balloon filled with sand.

At dinner that night, conversation turned to the adventures of the day:

> MCMENAMIN: We've reached that exalted state known as the cutting edge of science. We now know more about the Ediacaran biota than anyone else.
>
> SEILACHER: Yes, that's probably right.
>
> MCMENAMIN: It lasts about a week.

Dolf's key insight regarding the infaunal (in-the-sediment) nature of these fossils occurred while studying Dr. Pflug's collection in Lich, Germany. When he saw a cast of a double *Rangea* specimen he knew it couldn't have been at the surface. He had earlier reconstructed *Pteridinium* as partly emergent from the sediment surface, but now he thought that it was completely buried.

How did *Rangea* and *Pteridinium* secure food if they were immobile and lived beneath the sediment surface? Could they absorb food directly from pore water or the sediment itself? This might explain why ancient burrowers seem to avoid the Ediacarans; perhaps there is no food left dispersed through the sediments in their vicinity. Or, if they were indeed buried, were they simply trying to avoid desiccation in this tidal depositional environment?

Seilacher, in his counteractualistic fashion, thinks that conditions were different back then, perhaps with more food available within sediment. I am reminded of a quotation from L. P. Hartley: "The past is a foreign country; they do things differently there."

We saw several hunters earlier at Hellmut Erni's farm. They had shot an oryx. Dolf scornfully referred to them as neo-Nazis. And indeed, they did have German flag patches on their green fatigue-style jackets. Drunk and noisy at the hotel in Aus, they kept us up at night.

Our rooms were again cold for the night, and the air was so dry that Frieder and I put sunscreen cream on our faces for protection from chapping. We weren't bothered by bugs, however. The cleaning staff had sprayed insecticide (Doom Super) on our pillowcases.

August 7 was our shunpiking day, a day to turn off the main roads and go to Rosh Pinah. The road to Rosh Pinah would have been paved, but the lead-zinc and silver reserves in the mine are largely depleted, and in any case the price of lead has been low ever since it was taken out of the gasoline. We passed the Schutztruppe POW camp on the left, a dissolving ruins of mud brick buildings. Kubub Farm was on our left, and we were once again surrounded by smoothly rounded granite outcrops.

The Nama escarpment forms hats on the underlying granites. Deposition of the Nama Group was synonymous with what is called the Pan-African Orogeny, a lengthy mountain building episode, the greatest geological event of the continent. The supercontinent Rodinia was destroyed and the subsequent supercontinent Gondwana formed by this series of geologic events.

As we continued south to Rosh Pinah, Seilacher continued with his story about growing up in a German university town. The best known of these towns are Marburg, Tübingen, Göttingen, and Giessen. Dolf grew up in Tübingen, the world's leader in soap bubble production. He is now an emeritus member of the university faculty at Tübingen.

Early in his scientific career, Seilacher supported his fieldwork and research by collecting and selling mushrooms (white champignons). Dolf's favorite mushroom is the rock mushroom, or *steinpilz*. His expertise in field mycology is still remembered by elders of Tübingen, who periodically ask Edith Seilacher if she is the woman who married the handsome young mushroom vendor. Perhaps the entrepreneurship of young Seilacher's mycological fellowship later influenced his mushroom farmer hypothesis, which offers an explanation for otherwise problematic trace fossils. Seilacher has postulated that offshore burrow systems such as ichnogenus *Paleodictyon* (figure 4.6) represent sediment-walled microbial culture chambers, allowing the metazoan tracemaker to feast on the otherwise inaccessible banquet of refractory organic matter.

How were such mushroom farm burrow systems preserved? An early convert to the once radical concepts of density currents and turbidites, Seilacher now believes that the *Paleodictyon* burrow system is most often preserved at the base of a turbidite (submarine mudslide), where erosive scour followed by sand casting preserves the ichnofossil and protects it from obliteration by compaction of its muddy matrix. Not everyone agrees with Seilacher's assessment on this point, for some paleontologists believe that *Paleodictyon* is in fact a xenophyophore protist.[44]

We continued south toward Rosh Pinah, which is not far from the

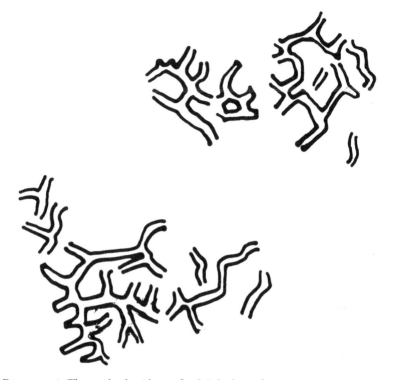

FIGURE 4.6: The graphoglyptid trace fossil *Paleodictyon* from the lower Cambrian of western Canada. Burrow geometry in many graphoglyptids is such that water flows continuously (by passive flow) through the interconnected passages.

Sketch from figure 3.10 of M. A. S. and D. L. S. McMenamin, *The Emergence of Animals: The Cambrian Breakthrough* (New York: Columbia University Press, 1990).

Orange River. We saw rare kokerboom trees and the red Nama sand dunes, then passed pre-Nama folded sediments of the Gariep Complex, contorted into recumbent folds. Soon we reached Rosh Pinah. Rosh Pinah airport. The landfill. Mine is on the left. Bougainvillea. Jacaranda trees. Rosh Pinah Drankwinkel. Rosh Pinah Bakkery. Rosh Pinah Winkel Staghuis. Volstruiss Strasse. Kokerboomstrasse. Ebbestrasse. Gemsbokstrasse. Ornamental junipers and a very large jacaranda. Tamarisk and thornless acacia. Monkey puzzle tree. Solar heat panels on the corrugated metal roofs of houses. Rosh Pinah is a mining town aspiring to be a solar-heated suburb.

Rosh Pinah mine is in the Rosh Pinah Formation at the base of the Gariep Complex. The mine tailings have been shaped into a gray ziggurat. The junkyard nearby is laden with cars. Heading out of town, we saw high-relief basin and range mountains, with scattered clumps of *Euphorbia*. I caught a large green *Agama* lizard at (not for) lunch. Delicate cream yellow flowers adorned either side of the road.

The Seilachers intended to take in the wildflowers and the local geology on this shunpiking day. We stopped over a valley rimmed by absurdly high-relief mountains of dark volcanics and metasediments of Gariep Complex. The rocks were tan and black near a vertical fault visible in the distance. We climbed up a slope with scattered *Euphorbia,* ice plant, and partially silt-covered botryoidal clumps of calcrete. The irregular calcrete lumps looked like multiple scoops of ice cream. Shells of deceased land snails, as thick-shelled as a marine moon snail, were scattered about in the dark silty soil. The shells had been sitting there a while, with the same furrow-and-rill, rain-etched weathering as we saw in weathered chunks of Schwartzrand Limestone at our first stop on the shunpike. I spotted another toad-shaped orthopteran nymph, this one nearly invisible, so closely did it match the color of the silt. After each jump these squat grasshopper nymphs land upside down, but they right themselves easily. A plant with low-spreading leaves looked as if it was lovingly manicured by a bonsai gardener. A green bottle brush plant was nearby.

White splotches on the dark outcrops looked like talc deposits but were actually exposed calcrete nodules. Apparently the local soil was being lost to erosion; no surprise on this steep and poorly vegetated slope. Continuing on, we passed schistose Gariep metamorphics, a bathhouse topped by solar panels, and a gravel road with white-knuckle blind curves.

We reached the banks of the Orange River and continued southeast. Although local rocks were still dark in color near the border with South Africa, we made a profound shift in geologic terrane. We had left the Precambrian and were in the Permian and Triassic rocks of the Karoo Sequence. Granite was on one side of the road, tillite (consolidated sediments left behind by melted glaciers) on the other.

A tall, solitary kokerboom tree with macelike branch terminations shaded the banks of the Orange along with the willowlike acacias. The Orange River floodplain, diamond conduit extraordinaire, was choked with diamond-bearing gravel flats and sand bars. A bluish gray Karoo outcrop jutted out of midstream. On the north bank, schists with pods of vein quartz were marbled with dark metallic gray mineralization. Bank flotsam included a rusted coil of barbed wire; Seilacher attributed this wire to the legacy of apartheid. A swallow flapped lazily over the outcrops on the banks. Other swallows milled around the river island outcrop, apparently a good spot for insects. A gravel bar downstream was stabilized by grass and small acacias. Near the water, the Karoo schist looked like weathered wood.

The Karoo tillite is called the Dwyka. The same series of late Paleo-zoic, glacially derived rocks crop out in South America, India, Australia,, and Antarctica. The Dwyka Tillite and its correlates on other continents were one of Alfred Wegener's best pieces of evidence favoring his theory of continental drift and the existence of an ancient southern supercontinent.

More recent researchers have tried to argue that the Dwyka and other ancient tills are not glacially derived at all, but are ejecta deposits hurled aloft by giant meteoritic impacts.[45] A few minutes on the outcrops dispel this notion. Even though they are pretty well metamorphosed here, the schists of the Dwyka show scattered, large rounded boulders stuck in the schist like plums in a pudding.

The boulders, up to a half meter in diameter, were unquestionably dropstones. Dropstones fall into marine silts when icebergs, calving off of the nearby glaciers, melt and release to the sea floor their suspended rocky loads. Dropstones are accidental "messages in a bottle," and the message to a geologist is always the same: A glacier was here.

I stood on a quartz-marbled outcrop overlooking the river and just north of the river island outcrop. This was the furthest south I had ever been. The Seilacher party drove me another half kilometer south, just for the fun of it, and I went no further south this trip.

Heading back, Frieder and I compared cameras. He preferred the metal body of my older Nikon FM-2 to the plastic body of his newer semiautomatic Nikon. While dusting his camera with a squeeze bulb brush, he joked with me that his camera was actually a firearm "easily converted to fully automatic."

The swallows were nesting in aeolian sand bluffs. The bluffs had been eroded to form a miniature Grand Canyon. A stately South African heron fished the river. Clasts in a tillite roadcut looked more angular than the ones in the river bank.

Ambitious off-road vehicles had left tracks on improbably steep silty slopes near the road—now that's real shunpiking! A curious sand apron abutted against dark tillite outcrops.

There was still no coffee available in Rosh Pinah, so we bought Vanilla First cookies and headed north. The dark hills on the receiving end of localized rain shadows had a green patina in the fading light. A painted sign on rock warned, "Speed kills also wear." Translucent-glass insulating discs ornamented the power lines in sets of six. Others, brown, looked like flying saucers. Spindle-shaped insulators stood atop poles like weaverbirds contemplating a new nesting site. Another sign read "Pyplyn."

We saw a conical hill of mass wasted rock debris that looked like a cinder cone but was not. Dolf Seilacher called it a geomorphic "pointlike singularity that makes a scree apron or Chinese hat." Edith Seilacher noted that we had been driving across a bajada for the last 35 km. Ahead, curving lenticular dikes in granite met the flat-lying Nama sediments. The dikes in the batholith looked like Hadrian's wall slicing up the slope.

We passed the Witputs Game Lodge. Oddly, I hadn't seen a single roadkill on this trip. Perhaps this was because of a combination of the rarity of traffic and game fences in most areas. Or did the jackals make fast work of any carcasses?

Furry weaverbird nests were common on the acacias as we reentered the Nama Plateau. The canyons and block faults of the Nama terrane were especially aesthetically pleasing, like the architecture of the Kyoto temple precinct. The outcrops were organized into four or five bedrock-controlled terraces, with kokerboom trees or other plants on the flat stretches between outcrop cliffs. The tall, black and tan cliffs might have invited monumental carvings in living stone, like Petra in Jordan, but instead we saw on the rocks an anomalous painting of the red and white Canadian maple leaf flag.

In the late afternoon light the Nama Plateaus to the north appeared smooth, like wooden armrests polished by generations of wear, not unlike the chairs near the fireplace of the Bahnhof Hotel in Aus. The comparison was apt, for the Namib desert is as much as 130 million years old,[46] and indeed the terrane gives evidence of millions of years of weathering in a desert environment. The desertification of this region began in the Cretaceous, as part of the climatic changes resulting from the breakup of Gondwana and the opening of the South Atlantic.

We passed the Aus marble quarry and several Rooibos trees, whose leaves and twigs are steeped to make tea. The dust from cars ahead drifted over the desert like a heavy mist in the sunset. To the east we had a clear view of the profound angular unconformity below the Nama. An unconformity similar to this, at Siccar Point in Scotland, led Hutton and Playfair to grasp the immensity of geologic time.[47]

August 8 was a Sunday and the slender, bright-eyed Bushman women served us plump pork sausages in addition to the standard, hearty breakfast fare of two eggs, two bacon strips, and three slices of tomato. The gas station in Keetmanshoop continued to advise motorists to "Fill Up and Feel Good."

After breakfast we drove toward the casting site. There was tension in the group, for this was the morning we were to learn whether the second attempt at casting was successful. The casting medium was Wacker Silicon (*kautschuk* in German), manufactured by Wacker-Chemie GmbH of Munich. The cold nights greatly delayed the hardening of the silicon. The Thursday before, with Erni's help, Seilacher's team built a wood frame and clear plastic sheet greenhouse to help warm the casting slab during the day and to preserve its warmth at night.

We arrived at the casting site at 10:20 A.M. Erni was already at the site, standing over the slab that Sunday morning as if in prayer. He was ready for the great unpeeling.

The first stage of the day's work was to make an achemie glue and plaster shape form over the silicon. This would preserve the overall shape of the slab, and the floppy silicon mold would be nested in it during the making of plaster and epoxy reproductions of the slab. When the shape form was finished and removed from the back of the silicon, it was finally the moment of truth.

The greenhouse-protected silicon had been left on the rock for several additional days to ensure proper curing. The visible surface seems firm enough but no one was sure whether it had hardened properly in the most important place—where it met the rock. It might just be a gloppy mess or otherwise poorly cured and useless. A ripple of relief moved through the group as a perfect silicon reproduction in negative was removed with great care from the rock slab.

Dolf puffed on a cigar as he launched into a lecture on *Ernietta* for Hellmut Erni. We were told that such cigars were a luxury reserved for holidays. This was especially so because he was recovering from major prostate cancer surgery. To the relief of all, the desert climate seemed to be hastening the healing of the surgical wound. Dolf chipped a piece of matrix off of an *Ernietta* specimen and then handed it to Frieder Pflüger to glue back on.

I collected up the three-dimensionally preserved *Pteridinium* specimens and carried them in a cardboard box back to the first site where we had attempted to excavate a slab. Dolf wanted to try to put these "rubble" specimens back into place to gain a better understanding of their original spatial relationships. He and Frieder vigorously brushed and swept the base of the bedrock puzzle, scraping at the calcrete like dentists removing calculus.

The end of the casting day was warm enough to bring out giant yellow biting flies, with sucking proboscises several millimeters in length. Fortunately, they were big enough and slow enough to not pose much of a threat to us.

That night, while we were having drinks in the bar (opened just for our group on Saturday), the Bottle Store attached to the Bahnhof Hotel was robbed of liquor. Apparently some of the unemployed Bushmen felt it was unfair that we were being served drinks on a Saturday while they were excluded, so they helped themselves. The generators stayed on well past 11:00 P.M. In the morning, the Bushman who usually sweeps the red steps in front of the hotel was refinishing it with what appeared to be reddish shoe polish. Unfortunately, his order to refinish was given several hours too late, for we kept walking on it to and fro from breakfast on Monday morning, August 9.

As a good-bye to our party, the cleaning and cooking ladies serenaded us with a beautiful South African song with pulsing, staggered harmonies. This was followed by a rousing German song.

We bade farewell to the Bahnhof Hotel and Aus and headed north toward Helmeringhausen and then Maltahöhe on minor routes C13 and C14, with a plan to skirt west and cross the border into the Namib desert (the Namib Naukluft Park). Soon we had a good view of red dunes to the east, and many ostriches on the east side of the road. The pale grasses looked to Frieder like "amber waves of grain." A solitary shepherd off to the right tended sheep. More ostriches (*Strauss* in German) and gemsbok. A pair of eagles with white bellies and dark upper surfaces glided above the town of Tirool (Afrikaans for Tyrol). Dark hills loomed on the left. We passed a date, orange, and *Opuntia* cactus farm on the right, and an odd radio receiving station with wires bent into two overlapping squares, offset by 45°, making it look like a rectangular star of David.

In the hills we turned west on D707. Vermilion red ridges of the Nagatis Formation of the Sinclair Sequence, 1200 million years old, appeared to the south. Soaring birds of prey, sheep, and windmills. The gravel farm roads made for hazardous driving, for the red dust really grabbed at the wheels.

Like a mirage in the desert, from the middle of nowhere and still 72 km from Maltahöhe, a stone castle appeared on the Duwisib farm, and we stopped for a visit. This was the Duwisib Castle of Hansheinrich von Wolf. Born in Dresden in 1873, von Wolf served in the Royal Saxon Artillery at Königsbrück near Dresden, and in 1904 volunteered for duty with the Schutztruppe after war broke out with the native Hereros in the South West Africa colony. He was awarded the Red Eagle Medal Class IV for his service. As Seilacher put it during a visit to a Schutztruppe cemetery on shunpiking day, "Medals are the stamps for stamp collecting in a deadly game."

In 1907, back in Dresden, Captain von Wolf married Jayta Humphrey, daughter of the American consul. Von Wolf was tall (1.98 m), energetic, and adventurous, and by all accounts a generous and hospitable sort, "whose attitudes roamed between reality and romanticism."[48] Mrs. Hoffman, wife of the chief Namibian Survey geologist, related a humorous aside: Whenever Captain von Wolf needed money he threatened to return to Germany, and his relatives dutifully complied.

The von Wolfs arrived in Windhoek in 1907 and inquired about farms for sale by the Treasury. They were recommended to the Maltahöhe district. The town Maltahöhe was founded in 1900 by district commissioner Hennig von Burgsdorf and was named after his wife, Malta.

Captain von Wolf purchased 140,000 ha in the district, at prices ranging from 30 to 80 pfennig per hectare. By 1911 the main residence at Duwisib, actually a castle, was complete. The residence is reputed to have cost a quarter of a million dollars to build. The plan of the castle is based on the enclosing of an inner courtyard, itself enclosed by an outer wall. It is well suited for defense, with corner risalites and battlements, and indeed gives the impression of a fortress. The castle, like many German colonial buildings in Namibia, is built in the neoromantic Wilhelminian style. In accordance with this style, the castle holds a portrait of Crown Prince Wilhelm. Duwisib Castle also includes some gothic and renaissance elements.

The castle, built of red sandstone, is richly ornamented with paintings, old furniture, copperplate engravings, photographs, and both authentic and ornamental weapons, most dating from the eighteenth and nineteenth centuries. The most interesting pieces of furniture are two wooden armchairs with Habsburgian double eagles, said to have belonged to King Philip of Spain in 1581.

Captain von Wolf was a passionate horseman, and numerous portraits of horses hang on the castle walls. He bred horses, including in his breedstock Afrikaner and Australian mares and Benito thoroughbred stallions.

In early 1911 von Wolf fell into financial difficulties and was unable to pay his obligations to the government. By 1913 he was threatened with court action if he failed to pay his tax debts. The debt was partly settled by government compensation for a Benito stallion that had died while von Wolf had lent it for stud.

In 1914 the von Wolfs left for England to buy more stallions. They were surprised en route to learn that war had broken out, and the ship veered across the Atlantic into a South American port. Captain von Wolf desperately wanted to reach Germany through the English blockade. After several episodes of intrigue on board a neutral ship, von Wolf and his associate, von Dewitz, finally snuck back into Germany. Captain von Wolf immediately reported for duty as an officer, and on September 4, 1916, was killed in France during the Battle of the Somme. Thus ended his romantic attempt to bring the knights of old to the new German colony. His wife, Jayta, who eventually returned to her parents' home in the United States, summarized her African sojourn as follows: "Oh, it was an interesting experiment."

The castle remains a beautiful but isolated attraction, and we sipped coffee next to the two huge jacaranda trees in the sunlit courtyard.

Heading north to Maltahöhe, we followed the edge of the Nama escarpment to the east. The road was made in massive concrete sections to protect it from gully-washing desert flash floods. We passed the fairly minimal dwelling of a Bushman family. The father wore a green jumpsuit and a colorful terrycloth hat. Three children played in the yard, and from the looks of the mother, two more were on the way. The family looked robust and healthy.

The escarpment to the east looked steep and bold and was colored in reds and an almost greenish blue hue where the slopes and strata supported vegetation. The slope rose steeply and steadily.

Springboks were bounding on the right. They made multiple leaps (Frieder says like a kangaroo) in tandem.

Moving away from the escarpment we encountered a Kori bustard, an erect bird standing a meter and a half in height, with a straight bill and a crest. It reminded me of the evolutionary tendency of some bird lineages toward gigantism. At a *T* intersection a sign read, "Solitaire via Zaris," which seemed appropriate: This is a fine part of the world if you value solitude.

Our discussion turned to water resources on the Aar farm. Hellmut Erni hit water at 20–60 m, but he also bored one dry hole to 250 m before giving up. Subsurface aquifers were of the bedrock joint and fracture variety. Drillers on the porous alluvial fans found more reliable water than those drilling directly on bedrock because alluvial fans are often superimposed over rangefront, bedrock-fracturing faults.

We passed by road cuts of thinly bedded sedimentary rocks. Many trees were adorned with small weaverbird nests, arrayed in ornament fashion. Several large nests were seen on telephone poles.

Maltahöhe appeared on the left, a collection of tidy-looking buildings in off-white with red and gray corrugated roofs. It did not look as if the night life would be particularly wild, although the tourist information center was enclosed in barbed wire. Standard Bank. Sonsky Modes. Maltahöhe Hotel: another one-star hotel.

The hotel had hot water, and in my opinion it deserved at least a star and a half. We began dinner at the hotel restaurant with drinks of Cardenal Mendoza and Johnny Walker. Over dinner, we had white wine, lager, and Rooibos tea. Some of us had an excellent roast pork dish called Schweinebraten. By the end of the meal some of our party showed signs of intoxication.

On August 10 we continued driving northward. A startled Kori bustard flapped away with a weighty motion. Gusty winds met us between Kalkrand and Rehoboth. The van stopped for photographs at the tropic of Capricorn, the same latitude as Rio de Janeiro. The last 10 km of the drive to Rehoboth looked comparatively lush. The acacias apparently had taproots sunk deeply into a ready source of ground water.

Old and new technologies collaborated in Rehoboth, as we encountered a horse-drawn auto trailer. Corrugated metal roofs were held down by stones. We had lunch at Sigi's a la Carte Restaurant. Music of the day was reggae country; we heard James Taylor's "How Sweet It Is" sung to a reggae beat. Dining in the restaurant with us was a white man with a black leather jacket and a slender black man with a narrow tie.

A dust storm rose in the east as we headed north; we were 80 km south of Windhoek. The minivan was not very stable as gusty winds buffeted the road; Hans had to wrestle with the steering column. The grasses were prostrate to the wind but, oddly, the tallest trees barely moved. The mountains to the west jutted boldly upward, giving an impression of the Alps. A green and white lorry, property of the F. D. du Toit Company, passed in the opposite lane. Could F. D. be a relative of Alex du Toit, the famous South African geologist (1878–1948, author of the famous book *Our Wandering Continents*)[49] and expert authority on the Gondwana supercontinent?

A green-blue bee-eater rested on the right side of the road. We returned to the staggered short-stick, jackal-proof fences. As we entered the foothills on the approach to Windhoek, the wind seemed to gain speed. It seemed to me as if this would be a stellar place for the development of wind-generated electricity farms, such as those at Altamont pass in California, although perhaps the wind gusts here would be too strong for this. On a different expedition I encountered just such a wind farm in southern California's San Gorgonio pass.

We saw a bold knob of granite on the right; a molar-shaped mountain on the right had similar knobs. A troop carrier passed in the opposite direction. Namibian soldiers stood in the back wearing green camouflage fatigues.

On the left, we passed a blasting area surrounded with barbed wire. The miners' huts had shallow-peaked cone-shaped metal roofs. The cylindrical huts were connected together in groups of three. On the right was a small cemetery. Growing in the cemetery was a large clump of cactus that looked just like the North American cholla. I knew from previous experience with this plant that it would be unwise to transplant it here. With 20 km to go to Windhoek, we saw a signal-repeating station high on a hill to the left. These high, rectangular structures appeared every 30 km or so.

The trees had regularly dendritic branches, an apparent adaptation to the wind. On shunpiking day (the Seilachers loved this term), we collected some of these fractal branches out of admiration for their morphology.

The Luipersvallei turnoff was on the right, and further on were hogbacks with bedded sediment on the right—Okahanja. 80 kph, speed enforced by camera. Stadium on right. Ero recreation area, Jeans Street, then back to Hotel Safari. Ontvangs/Receptions/Anmeldung. Utility truck with ladder on top: Telecom Namibia. S. W. A. Safaris (Pty) Ltd. Windhoek green and white tour bus. Handsome, confident-looking Ovambo bellhops, with black pants and shoes and blue or red button-up shirts with epaulets. Black men in suits and narrow ties. One drove away in a BMW. Hotel Safari (III TYYY). White woman with character and a confident look.

Guthenbergstrasse. Faradaystrasse. Benzstrasse. Benz built the first German car, and his wife was the first woman driver. Gallagher Fence Systems. Swatyre (Tire Store). AUBob Funeral Service. Wecke and Voígts. Liquor store ad for Sedgwick's The Original Old Brown Sherry. The Weavers Nest. Swaco House. Swafo Travel Agency. The names of several businesses sound something like SWAPO. There was a street called Macadam Street; difficult to argue with that. A Robert Redford lookalike in a poster picking a nylon string guitar: "Grooving in the 60's . . . still grooving today." At the corner of Rehobotherweg and Nachtigelstrasse we picked up cardboard boxes with bottle spacers for the packing of rocks.

The next day we again arrived at the Namibian Geologic Survey for an extended visit with chief geologist K. H. Hoffman. We also spoke with economic geologist Niall McManus, whose current project was to map the mineral occurrences in Namibia. He noted that the Japanese Metal Agency was prospecting in Namibia for the alkali metal deposits associated with anorogenic ring complexes. It is illegal in Namibia to own uncut diamonds.

McManus has a bachelor's degree from Trinity College in Dublin and a master's from Imperial College. He lived in Boston for 3 months, working for Greenpeace, and later held a job in Manhattan. He especially enjoyed Manhattan, where he "worked for dollars rather than for compassion."

McManus once pulverized a black mineral known as Bushveld chromite, mixed the powder with epoxy, and poured the mixture into a mold of a trilobite fossil to make an attractive reproduction. A distinguished colleague, noting the cast on McManus's desk, picked it up and correctly identified both the tiny fragments of chromite and the genus of trilobite, but failed to notice the compositional disparity. Chromite is a mineral found only in fossil-barren igneous rocks.

In the Survey laboratory we met with the grind of rock saws, the low thump of the fume hood fan, and the smell of acetone. "Namibia Land of the Brave" hangs on the bulletin board. A poster proclaiming "Our Namibian Heritage Meteorites Are Protected by Law." Salpetersuur/Nitric Acid. A Frantz Isodynamic Separator Model L-1. Soutsuur/Hydrochloric Acid.

At the end of the hall in the Survey building hung a field photograph of geologist Henno Martin, holding a gnarled walking stick, his appearance gaunt as he gazed off into the horizon. He looked like a white bushman. We encounter Martin again in Chapter 13 of this book.

We had spent the evening in the home of a Dr. W. Hegenberger. His house was up on a hill and had a splendid view of the city. We had an enjoyable time with Dr. Hegenberger and his talkative and gregarious German-speaking relatives.

Security measures are essential in Windhoek. Razor wire (optimized for slicing human flesh) and barbed wire surround the Geological Survey building to deter theft. Hegenberger's home had barbed wire as well.

Chief geologist Hoffman drove a 320L BMW. In the Survey parking lot I saw Toyotas, Golfs, Citigolfs, Monzas, Mazdas, Isuzus, and a Nissan Safari 4×4. The Volkswagen Beetle lived on in health in Namibia. Not an American-made vehicle in sight.

We left the Survey and headed back into town. Oka Puka Sand Curt-Von-François dumptruck. Four closely spaced, young Canary Island palm trees. Slowly moving blind black man, cane held far out in front. Violets. Bougainvillea. Kalahari Sands Hotel. The Eros airport was very busy with propeller planes. Parking meter was "kaput," saving us change.

The center of downtown Windhoek is a plaza filled with colorful, Disneyesque storefronts and kiosks. A Namibian mall. "Namibian Career and Manpower Consultants." Jive—The Cool Cooldrink. Le Bistro corner cafe—cappuccino and goulash soup. Blind man reading from a Braille Bible in an African dialect. Urbane, well-fed Bushmen and Bushladies. T-shirt: "Spoil Sport." Sign:

AVIS
We try harder
AT EROS AIRPORT
Tel. 33166 A/H Telepage 52222

Woman carrying oranges on head. Black woman smiled and waved at me. Seilacher thought that she just liked my hat, a desert-style pith helmet. The Germans were delighted when I first put it on, calling it my "tropical helmet."

Buxmann, The Professional Furnishers. Nama Craft Store-a Newveld enterprise. As per my introduction to this art at the Frankfurt Airport, the shape of the wood controls the carving of wood sculptures in Namibia. The sculptures appear misshapen until you are familiarized with the style. And then one sees something akin to Picasso. Or perhaps a !Xam San rain creature.

At the very center of the middle of Windhoek, at the epicenter of the nation itself, was (other than diamonds) the greatest natural national treasure: the Gibeon meteorite fall (figure 4.7). Recall the prominent poster in the Survey that warned against illegal removal of meteorites from Namibia; Namibians were still smarting from the loss of much of the huge Gibeon meteorite fall to international dealers.

The meteorites of downtown Windhoek were set on harmonically stepped granite pedestals, composing a striking statuary that appeared to sweep down from the sky. One of the meteorites in the display was sliced and polished to show the angular crystalline Widmanstätten structure. From the brass plaque:

The largest known meteorite shower to fall to earth covered an area 360 kilometers long and 100 kilometers wide around Brukkaros. Most fragments fell just southeast of Gibeon. [Gibeon was founded by Hottentot settlers who, under their chief, Moses Witbooi (father

FIGURE 4.7: The Gibeon meteorite display in downtown Windhoek, Namibia. Peter Seilacher and the author stand beside a pedestal-mounted meteor.

of Hendrik Witbooi), named the settlement after the scene in the Old Testament where Joshua calls upon the sun to stand still as he avenges himself on the Amorites.] The explorer, J. E. Alexander, recorded the occurrences of the meteorites in 1838, although they had long been known to the local Namas who hammered pieces into implements.

A total of 77 pieces have been found having almost identical chemical compositions; these are believed to have initially been part of one large body over fifteen tons in weight which fragmented long before its individual pieces entered the Earth's atmosphere. The largest fragment found weighs 650 kilograms.

The meteorites are classified as octahedrites, the most common type of iron meteorites, and consist entirely of taenite and kamacite, two different crystalline phases of iron-nickel alloy, the former containing much more nickel than the latter. These two phases form alternating parallel crystal bands that are arranged in a triangular pattern referred to as Widmanstätten Structure [figure 4.8], a characteristic of all octahedrites. Besides Iron, the meteorites contain an average of 8% Nickel, 0.5% Cobalt, 0.04% Phosphorus, small

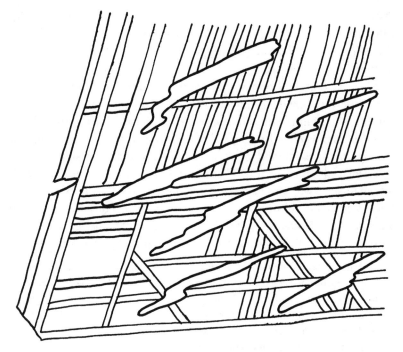

FIGURE 4.8: Widmanstätten Structure of the Gibeon meteorite formed by alternating parallel crystal bands that are arranged in a triangular pattern. Width of rock illustrated 4 cm.

amounts of Carbon, Sulfur, Chromium and Copper, and traces of Zinc, Gallium, Germanium and Iridium.

Between 1911 and 1913, Dr. Paul Range [for whom *Rangea* is named], state geologist for the German South West Africa government, collected 37 fragments. Several specimens have been donated to research institutions all over the world and 33, with masses ranging from 175 kilograms to 555 kilograms, remain in Windhoek today.

National Monuments Council

August 12, 1993: Having left Africa and returned to Germany, Seilacher and I are meeting with Prof. Hans D. Pflug at his home in Lich, Germany.

Pflug is a pleasant and soft-spoken man. He welcomed me warmly into his home, said he was very pleased to meet me, and handed me a copy of an article I had written that had been translated into German.[50] Pflug's mind is abroil with scientific hypotheses, many of them unorthodox. In one of his recent articles he argues that differences in the early earth's gravitational field prevented animals from reaching large size in the past.[51] To say the least, this hypothesis will be met with skepticism by most other scientists.

Also a veteran of World War II, Pflug served in the German navy on the battleship *Tirpitz*. Pflug was wounded during a short post-Navy career as an infantryman. He showed me both snapshots of the *Tirpitz* and the scars where a Russian bullet had entered near his wrist, passed through his forearm, and exited near his elbow. Only a flesh wound.

Displayed before us was Pflug's entire collection of Namibian Ediacaran fossils. I wouldn't have thought it possible, but the diversity of form and preservation dwarfed anything in Erni's shed. Pflug was originally trained as a paleobotanist, and this may explain why he, in the 1960s, made some unusually important observations regarding the Nama fossils.

> Seilacher began by reminding me that Professor Pflug was the first person not to use the Glaessnerian shoehorning, in other words, the shoehorning of Ediacaran taxa into modern taxa such as the phyla Cnidaria or Arthropoda. Seilacher graciously acknowledged that he had learned this first from Pflug, and that Pflug had made the extremely important observation that Ediacarans represented a completely different lineage, and were perhaps even prokaryotic.
>
> Later discussion turned to the specimen of *Rangea* shown in figure 4.9. Seilacher asked Pflug whether we could discuss preservation of the fossils, considering the uncertainties surrounding this issue. Pflug asked which aspects of preservation he would like to discuss. Seilacher said that he wanted to learn more about the preservational conditions. It was clear to Seilacher, from his observations of the fossil localities in Australia and Siberia, that the preservational agent was smothering, obrution,[52] or "Verschuettung,"[53] and that in these depositional environments the most likely cause of such burial was storms. Storm sedimentation covered the fossils and preserved them as a relief on the bedding plane—"Pompeii in a storm sense."
>
> At Mistaken Point in Newfoundland, Seilacher saw the same sort of thing. He had counted 15 different stratigraphic horizons in which the smothering had been accomplished by volcanic ash. Each horizon is a layer of graded tephra or volcanic ash particles. The lowermost layer is prominent as a white band, and it is at the base of this white band that the fossils are preserved. As in Pompeii, the fossils were smothered exactly in living position. They were covered by volcanic ash, and the only generations of organisms to preserve as fossils were those that were alive on each of the 15 successive "Pompeii days." Other generations were certainly present but are not represented by fossils in the intervening strata. The fossils can preserve only at the interface between the upper, muddier phase of the lower ash layer (on which the organisms were living) and the newer, falling ash that snuffed out the community.

FIGURE 4.9: *Rangea schneiderhoehni* from Namibia, specimen in the collection of H. D. Pflug. Shown are four subfronds converging at the tip of the composite frond. Scale bar in centimeters.

I added that this was similar to the way in which fossils preserve at the base of a turbidite layer. Exactly so, Seilacher replied. He noted that there is an important difference between the fossils in Newfoundland and Australia. Fossil relief is opposite in the two localities. In Australia the fossils are in negative relief on the base of the smothering bed; in Newfoundland they are in positive relief on the top of the smothered bed.

Later discussion turned to "deformed" specimens of *Pteridinium* in the Pflug collection. Seilacher noted a specimen of *Pteridinium* that, in his view, grew upward, turned around, and changed polarity (what had been the underside of the "bathtubs" became the insides of the "bathtubs"). Seilacher insisted that this indicated that the ventral and dorsal sides of the organism have no significance in the ordinary, animalian sense. What was ventral here becomes dorsal there. I added that this was the firmest proof we have ever had that *Pteridinium* is not an animal. Seilacher asked Pflug whether Hans would be allowed to make a cast of this specimen.

In later discussion, Seilacher mentioned to Pflug and me that in the Namibian discussion, Runnegar and Seilacher had come close to agreement on a number of points. In particular, Runnegar seemed to agree that forms such as *Pteridinium* did not fit any known metazoan body plan, and could be neither ancestral to nor related to any modern body plans. Pflug agreed with this. Seilacher hastened to add, however, that the taxonomic placement of *Spriggina* and *Dickinsonia* were not agreed upon in Aus. Runnegar felt that these forms were unrelated to *Pteridinium,* and that their resemblances were superficial, a consequence of convergent evolution resulting from shared evolutionary responses to a unique moment of Earth history.

The hard-won silicon and its shape form remained in Germany for some time for casting at Tübingen, but Seilacher eventually brought it with him to Yale University. There his technician made three casts, one in plaster and two in fiberglass. The fiberglass casts were painted an even tan color to approximate the color of the Nama sandstones. I received one of the fiberglass casts and carried it in a geology van back to Mount Holyoke College. The cast was a marvelous creation and, in my possession, immediately led to long-lasting study sessions with my paleontology students.

In fall of 1993 I had the honor of presenting Dolf with the Medal of the Paleontological Society. On ascending the platform to receive the award, he fell off the podium, but fortunately was not seriously injured. I mentioned the slab in my citation for him, noting that it was "arguably the best contiguous slab of Ediacaran fossils in the world." When my citation for Seilacher was published in July 1994, it apparently caught the attention of prominent paleontologists.[54]

In mid-August I received a telephone call from the office of Stephen Jay Gould at Harvard University, asking whether I would provide an illustration of the slab for an article he was writing for *Scientific American.* I agreed to do this and asked for someone to contact me with more details.[55] A good deal later I was contacted by Michelle Press at *Scientific American* with a second request to do a photograph for them. She said she wanted a color photograph, and that she wanted it within the next couple of days. Regarding the photograph's quality, she said, "It has to be stunning."

This posed a problem for me, as my copy of the slab was not properly prepared for the making of a stunning photograph. The tan paint on the surface was attractive but lacked sufficient contrast. I needed some way to boost the contrast of the specimen and I needed it done quickly.

I raced home to my house and found a small can of dark wood stain and a brush. Going back to campus, I tested the stain on a hidden part of the slab and test results were encouraging. So I painted the entire slab with a thin coat of wood stain.

The process worked well. When finished I had a slab with sufficient contrast. It looked like "natural" rock but was much more attractive than the original surface of casted rock surface had ever been. I carried the slab cast three floors downstairs from my office to the back of Clapp Laboratory and set it up outside because I felt that natural lighting would be best for the photograph. Rain was threatening but the diffuse light through heavy clouds seemed to provide the lighting I required. I shot an entire roll of color film, and raced into Springfield, Massachusetts, for one-hour developing.

Results were pleasing (color plate 12). Michelle Press said that the photos were indeed stunning and looked three-dimensional, and said, "We are very grateful to your for your heroic efforts on our behalf." The result, which appeared as a full-page illustration to lead off Gould's article, appeared in the October 1994 special issue of *Scientific American* titled "Life in the Universe."

Seilacher seemed quite pleased at this accomplishment, only the first of what I anticipate will be many more fortunate results of his eclectic expedition to Namibia.

Notes

1. J. K. Wright, "Foreword," in C. H. Hapgood, *Maps of the Ancient Sea Kings,* pp. ix–x (Philadelphia: Chilton Books, 1966).

2. The name *Seilacher* is derived from *Ache,* ancient German for "river," and *Seil,* meaning "rope," although the latter may be a corruption (A. Fischer, written communication, 1996). Thus the name may mean "meandering river." Seilacher himself puns that the name is akin to selachians, the sharks.

3. See p. 325 in A. Seilacher, "Divaricate Patterns in Pelecypod Shells," *Lethaia* 5 (1972):325–343.

4. Incidentally, this is where Seilacher celebrated his official retirement on October 1, 1990, at 0.00 hours.

5. A large body of granitic rock, formed by cooling lava below the earth's surface.

6. Eldridge M. Moores, personal communication, March 2, 1997. Moores spent most of a week with Martin doing field work in the vicinity of the Orange River, Namibia.

7. *Lithops* seeds can be purchased for $3.99 from Thompson & Morgan Inc., Dept. RT, Jackson, New Jersey 08527. Ask for "Living Stones," Lithops, 2546 Greenhouse Perennial, net weight 4 mg. The instructions state that "no water should be given at all throughout the winter and early spring."

8. J. D. Ward and I. Corbett, "Towards an Age for the Namib," in M. K. Seely, ed., *Namib Ecology: 25 Years of Namib Research,* pp. 17–26 (Pretoria, South Africa: Transvaal Museum, 1990).

9. Pronounced VIND-hook.

10. E. M. Thomas, *The Harmless People* (New York: Knopf, 1959).

11. R. M. Miller, "A Note on Three Unusual Sedimentary Structures in

Sandstone of the Auborus Formation, South West Africa," *Journal of Sedimentary Petrology* 45, no. 1 (1975):113–114; see also A. Seilacher, *Fossile Kunst: Albumblätter der Erdgeschichte* (Korb, Germany: Goldschneck-Verlag, 1995).

12. Dacqué served as Heinz Lowenstam's doctoral adviser.

13. See p. 28 in P. A. Munz, *California Desert Wildflowers* (Berkeley: University of California Press, 1962).

14. See p. 13 in R. Williams, *The Non-Designer's Design Book* (Berkeley, California: Peachpit Press, 1994).

15. An unconformity is a level in a sequence of layered rocks where there was a significant break in the deposition of sediments. If the rocks below this level have been tilted and eroded, it is called an angular unconformity.

16. J. H. Wellington, *South West Africa and Its Human Issues* (Oxford: Clarendon Press, 1967).

17. A. J. Tankard, M. P. A. Jackson, K. A. Eriksson, D. K. Hobday, D. R. Hunter, and W. E. L. Minter, *Crustal Evolution of Southern Africa* (New York: Springer Verlag, 1982); G. J. B. Germs, "Implications of a Sedimentary Facies and Depositional Environmental Analysis of the Nama Group in South West Africa/Namibia," *Special Publications of the Geological Society of South Africa* 11 (1983):89–114.

18. Which, according to Seilacher, all have their convex sides down. This would have been an unstable position had they been transported.

19. Indicative of Dutch culinary influence.

20. Float specimens are rocks that have been broken off of, but have not traveled far from, their bedrock source.

21. Seilacher has interpreted *Protolyellia* as both a psammocoral (A. Seilacher and R. Goldring, "Class Psammocorallia (Coelenterata, Vendian-Ordovician): Recognition, Systematics, and Distribution," *Geologiska Föreningens i Stockholm Förhandlingar* 118 [1996]:207–216) and as probable microbial sand balls (written communication, March 3, 1997).

22. These men belonged to the !Xam people of the San (Bushman) language group. The ! refers to the click sound of this language.

23. W. H. I. Bleek, L. C. Lloyd, and G. M. Theal, *Specimens of Bushman Folklore* (London: George Allen, 1911).

24. A. Solomon, "Rock Art in Southern Africa," *Scientific American* 275 (1996):106–113.

25. I. Tattersall, *The Last Neanderthal* (New York: Macmillan, 1995).

26. Anonymous, "Namibia's Ancient Big Bird," *Science* 263 (1994):176.

27. Thrombolites are stromatolite-like structures lacking the distinctive internal laminations or layering.

28. Brain now says that he has found fossils in these cherts.

29. H. D. Pflug, "Neue Fossilreste aus den Nama-Schichten in Südwest-Afrika," *Paläontologische Zeitschrift* 40 (1966):14–25.

30. J. W. Valentine, "*Dickinsonia* as a Polypoid Organism," *Paleobiology* 18 (1992):378–382.

31. The tube worm fossils associated with Ediacarans in Australia; see B. Runnegar, "Proterozoic Eukaryotes: Evidence from Biology and Geology," in

S. Bengtson, ed., *Early Life on Earth,* pp. 287–297 (New York: Columbia University Press, 1994).

32. Easternmost locality (G) in Pflug's *Ernietta* paper; H. D. Pflug, "Neue Fossilreste aus den Nama-Schichten in Südwest-Afrika," *Paläontologische Zeitschrift* 40 (1966):14–25.

33. What Seilacher called "Flädle" structure (after a Swabian pasta dish) in the Kliphoek member of the Nama is what is commonly known as "dish structure," formed by intrastratal dewatering within a mature and well-sorted bed of sand (Frieder Pflüger, personal communication, September 13, 1996).

34. G. R. McLachan, "Taxonomy of *Agama hispida* (Sauria: Agamidae) in Southern Africa," *Cimbebasia, Series A,* 5, no. 6 (1981):219–227.

35. See p. 147 in R. J. F. Jenkins, "Functional and Ecological Aspects of Ediacaran Assemblages," in J. H. Lipps and P. W. Signor, eds., *Origin and Early Evolution of the Metazoa,* pp. 131–176 (New York: Plenum, 1992).

36. M. McMenamin, "Burros," *The Stanford Daily* 171, no. 41 (1977):2; S. Stueckle, "Mustang Savvy," *Hemispheres* (June 1994):80–89.

37. New York: Random House, 1990.

38. J. H. Wellington, *South West Africa and Its Human Issues* (Oxford: Clarendon Press, 1967).

39. M. McMenamin, *Carthaginian Cartography: A Stylized Exergue Map* (South Hadley, Mass.: Meanma Press, 1996).

40. J. S. Malan, "Social Evolution Among the Ovambo," *Cimbebasia, Series B,* 2 (1978):12.

41. In an early (and, as it turned out, incorrect) application of continental drift theory, Merensky hypothesized that the Namibian diamonds were a result of the release of deep-seated (and thus diamond-bearing) lavas, erupted along the rift canyon separating South America from western Africa. In fact, the diamonds derived from inland kimberlite deposits are not directly associated with continental breakup.

42. A rock type of explosive volcanic origin with a heterogenous fine-grained matrix; often contains diamonds.

43. R. M. Miller, *Namibia Geological Map, 1:1,000,000* (Windhoek: Geological Survey of Namibia, 1990).

44. D. D. Swinbanks, "*Paleodictyon:* The Traces of Infaunal Xenophyophores?" *Science* 218 (1982):47–49; L. A. Levin, "Paleoecology and Ecology of Xenophyophores," *Palaios* 9 (1994):32–41.

45. M. R. Rampino, "Tillites, Diamictites, and Ballistic Ejecta of Large Impacts," *Journal of Geology* 102 (1994):439–456.

46. J. D. Ward and I. Corbett, "Towards an Age for the Namib," in M. K. Seely, ed., *Namib Ecology: 25 Years of Namib Research,* pp. 17–26 (Pretoria: Transvaal Museum, 1990).

47. I'd like to see someone publish a picture book, a photographic essay of all the world's great angular unconformities.

48. See p. 20 in N. Mossolow, *Hansheinrich von Wolf und Schloß Duwisib [Hansheinrich von Wolf and Duwisib Castle]* (Swakopmund, Namibia: Society for Scientific Development, 1992).

49. A. du Toit, *Our Wandering Continents* (Edinburgh: Oliver and Boyd, 1937).

50. M. A. S. McMenamin, "Das Erscheinen der Tierwelt," in H. D. Pflug, ed., *Fossilien: Bilder frühen Lebens,* pp. 56–64 (Heidelberg: Spektrum-der-Wissenschaft-Verlagsgesellschaft, 1989). This book is sold with a separate geological time scale insert meant to be used as a bookmark—splendid idea!

51. H. D. Pflug, "Role of Size Increase in Precambrian Organismic Evolution," *Neues Jahrbuch für Geologie und Paläontologie, Abhandlungen* 193, no. 2 (1994): 245–286.

52. From Latin *obruere,* which means "to uproot or excavate."

53. A term commonly used in conjunction with snow avalanches in the Alps smothering a group of skiers (F. Pflüger, personal communication).

54. M. A. S. McMenamin, "Presentation of Paleontological Society Medal to Adolf Seilacher," *Journal of Paleontology* 68 (1994):916–917.

55. I had done work before for this magazine; M. A. S. McMenamin, "The Emergence of Animals," *Scientific American* 255 (1987):94–102.

5 · Back to the Garden

These organisms are adapted to the capture and full use of feeble
luminous radiations.   —Vladimir I. Vernadsky[1]

Where there is one photon of light, there will be an organism there
to capture it.   —Lynn Margulis, 1995

In 1984 I accepted a position as assistant professor of geology at Mount
Holyoke College in Massachusetts. I was very pleased to have been
offered this position, but it was quite a shock to move to New England
with my wife Dianna and our two-year-old daughter from balmy Santa
Barbara, where I had just finished graduate school at the University of
California. Without a direct maritime influence to moderate the climate, winter in western Massachusetts is lived close to the elements. As
I write in January 1996, the snow is falling heavily, and there are already
3 ft of snow on the ground.

For erstwhile Californians the transition to New England is highly
stimulating. At Mount Holyoke I found myself in the midst of a roiling
intellectual environment. I was infused with a newfound sense of community shared during the long winters.

While I was doing some background reading one cold winter's
evening, I came across Stewart Brand's book review of Lewis Thomas's
influential book *The Lives of a Cell*.[2] Brand describes "the symbiotic theory of Lynn Margulis" and the process of symbiogenesis by which, "in the
course of deepening cooperation," organisms evolved into subtly complex
forms. Brand noted that "Thomas makes you care about such things."

After reading the review I read Thomas's book. After reading *The
Lives of a Cell* I became intrigued by the possibility that symbiosis had
played an important role in the evolution of not only the eukaryotic
cell, but also of larger organisms. I had been aware of Margulis's serial
endosymbiotic theory since the early 1970s, when, as a junior high
school student, I had corresponded about the subject with my botanist
uncle, Joe McMenamin.[3] Later, in the early 1980s, I met Margulis at
the University of California at Santa Barbara.

The application of symbiogenesis to problems of the Precambrian-
Cambrian transition first came to my mind while I was teaching my

course Geology 341, "Great Ideas in Geology." As a new professor, I only had three students in the class, but it was an inspired group. We spent quite a bit of time in this course discussing the Ediacaran biota, and in fall 1984 one of the students in the seminar, commenting on the fossil *Charnia* (color plate 2), mentioned that it certainly looked like a leaf impression and asked how we can be sure that it is not a plant. We talked about this in the class, noting that vascular plants did not evolve until hundreds of millions of years after the apparent demise of the Ediacaran biota, the implication being that *Charnia* could not by any stretch of the imagination be a chestnut leaf-the similarities must be superficial.

I thought about these issues further, thinking back to a course I had taken with biologist Robert Trench at Santa Barbara. Trench is an authority on coral-microbe symbiosis, and he knows a lot about the importance of the association for corals. As far back as 1965, paleontologists had begun to recognize the features of photosymbiosis and even chemosymbiosis in Ediacarans. As a result of this training in graduate school, and the class discussions as a professor teaching his first seminar course, it dawned on me that many of the Ediacarans, perhaps *all* of the Ediacarans, appeared to be modified for photosymbiosis and light collection.

I took no action on this thought until the next semester. During January break I noted with great interest a very brief mention of photosymbiosis in Seilacher's 1984 paper on the Ediacarans.[4] But the breakthrough for me in this line of research came in spring 1985 as I sat doing research in my windowless office in Williston Library at Mount Holyoke. I was reading a fairly well-known paper by Al Fischer, a professor of geology at the University of Southern California, and came across the following comment hidden in the body of the text:

> The first break in this pattern of isolation came when some animals took photosynthetic algae into their tissues, in zoöchlorellar or zoöxanthellar symbiosis. . . . Hedgpeth [1965] has pointed out that some members of the Ediacara fauna may have lived in this fashion, for Glaessner [1962] reports that the jellyfishes lie in an "upside-down," that is, mouth-up, position, which is the normal one for the living, zoöxanthella-dependent jellyfish *Cassiopeia*.[5]

Fischer's scientific insight lay dormant for many years. Although J. William Schopf, Bruce N. Haugh, Ralph E. Molnar, and Donna F. Satterthwait note that "autotrophs would be expected to have preceded eumetazoans[6] in colonization of the benthic environment," they infer that these early autotrophs were thallophyte algae and apparently missed Fischer's insight concerning the Ediacarans.[7]

It is instructive to examine the Hedgpeth and Glaessner citations mentioned by Fischer. Glaessner does indeed talk about Ediacaran medusoids as upside-down jellyfish:

> It was considered peculiar that these fossils are generally preserved with their convexity directed downwards while dead jellyfish are most found on present-day beaches with their dorsal side upward. This is, however, not a serious obstacle to considering the fossils as medusae, since some medusae, at least, come to rest on the sea floor with their convex side downward.[8]

However, Glaessner does not make any paleoecological inferences based on this position. Instead, he rather pointedly avoids any paleoecological interpretations of the Ediacarans: "It is not my intention to pursue the discussion on palaeo-ecology further, pending completion of field and morphological studies."[9]

This sentiment is a reflection of Glaessner's natural caution as a scientist, his distaste for publication of premature results,[10] and perhaps fear that his colleagues would look with disfavor on any mention of endosymbiosis. I have a direct acquaintance with Glaessner's cautionary (as well as diplomatic) side, for the only time I corresponded with him he had this to say:[11]

> Thank you for your letter of 1 November and enclosures. I had previously known only your published abstract.[12] Pressure of work does not permit me at this moment to comment adequately on your interesting hypothesis but I hope to do that later. However, my first reaction to it is that it can be tested adequately only when the geophysicists make conclusive studies of palaeomagnetism during Precambrian-Cambrian time and specialists in tectonics decide where the Late Precambrian plate boundaries can be found. The relative positions of the Ediacaran localities around the equator being still unknown and all possibilities being advocated by different specialists, a palaeobiogeographic hypothesis seems to me somewhat premature (but worthwhile).

Unfortunately, Glaessner died before I had a chance to meet him.[13]

Joel W. Hedgpeth hints at a connection between Ediacaran fossils and a cnidarian known to be photosymbiotic: "Glaessner's description of the upside-down manner of fossilization of some of the medusoids suggests that they may have been similar in habit to the sedentary rhizostome *Cassiopeia.*"[14]

Hedgpeth does not mention photosymbiosis. He might have been thinking about it, but because he doesn't mention it ("habit" could merely refer to the jellyfish's preferred orientation), the credit for the photosymbiosis insight goes to Fischer.[15] Later in the article Hedgpeth does take a stab (and a quite accurate one at that) at the paleoecology question: "While obviously not a simple system, since the variety of fossils suggests a fair range of niches, the system represented by the Ediacara fossils lacks apparent predators and organisms with heavy shells."[16]

The only other suggestion of a linkage between the origin of animals and photosynthesis was made in 1953 by A. C. Hardy at Oxford University. Hardy argued that animals (metazoa) evolved from "unspecialized and relatively simple Metaphyta."[17] Hardy also states,

> The many different devices evolved by the various carnivorous plants to enable them to secure their food suggest how such a metazoan organism may have been derived. . . .
>
> The gradual transition from a simple metaphyte to a simple polyplike metazoan—a bladderlike cavity with tentacles—seems no more difficult to conceive than the evolution of the higher animal-like insectivorous plants; we have only to imagine the process going so far as to cut out all photosynthesis, thus making the organism holozoic as indeed has occurred repeatedly at the unicellular level of the flagellates and algae.

Trevor D. Ford, in his 1958 description of *Charnia* as an alga, had implied that the organism was photosynthetic.[18] I later asked Fischer whether the Ediacaran photosymbiosis idea really was original with him. He replied, "Yes, so far as I know that thought of photosymbiont associations was original with me—not that that seems so important. I find it hard to think of primitive things like that getting so big without some such mechanism—and I had no idea then of how big *Dickinsonia* actually gets or got."[19]

Mikhail A. Fedonkin was thinking along similar lines when he published an article in Russian[20] titled "The Ecology of Precambrian Metazoa of the White Sea Biota." On p. 29 he makes the case for photosymbiosis:

> Extant cnidarians, turbellarians, and representatives of other invertebrate groups are known to have symbiotic algae within their tis-

sues. These algae participate in oxygen exchange with, and provide much of the food requirements for, their animal hosts. It is possible that this type of symbiosis is one of the most ancient, and furthermore it is possible that it was significantly more widespread in the Vendian than it is today. If this is so, then the distribution of the flattened forms of Vendian Metazoa in shallow-water habitats conducive to the collection of the greatest possible quantities of light makes sense since such habitats would be essential for the light-intensive metabolism of the symbiotic algae.[21]

Writing in 1983, Fedonkin was apparently unaware of Fischer's work. Considering the great tradition of symbiogenesis research in Russia, I am surprised that Fedonkin did not go further with the idea of Ediacaran photosymbiosis.[22] He surely must have been exposed to the idea of symbiogenesis early on while a Russian student in the natural sciences. Later, Fedonkin attends to the idea of Ediacaran photosymbiosis at greater length after learning of my Garden of Ediacara theory (see below, where Fedonkin finally cites Fischer).

With Fischer's mention of photosymbiosis I felt that I had correctly established the pedigree of an important idea; such considerations may seem overly academic, but they are important in science for establishing priority, revealing the context in which scientific advances are made, and, by no means least important, revealing who one's potential scientific allies might be. The latter is particularly important for scientists who, as I was in 1985, are still in the early stages of building a professional career.

Emboldened by my understanding of the history of the problem, but still not fully confident of my ability to convince skeptical colleagues about Ediacaran symbiosis, I made the following addendum to my abstract for the Annual Geological Society of America meeting in Orlando, Florida: "The energetics of photosynthetic endosymbiosis are favored when the supply of nutrients is limited."[23]

I based this comment, penned in July 1985, on Pamela Hallock's work on the energetics of the host-zooxanthellae association.[24] The real implications of the idea did not hit home until only minutes before I was to give my 15-minute talk at the meeting. Sitting in my motel room in Orlando, I was reviewing my notes before taking the short walk to the convention center. I had a sudden and well-timed insight: If photosynthetic photosymbiosis characterizes the Ediacaran creatures, could this be a general characteristic of Proterozoic ecosystems? Could we be dealing with a predator-free ecosystem? A peaceful seafloor garden? A Garden of, of course! Ediacara.

I used this line at the end of my talk, and the audience found it quite amusing. More important, it got my point across very clearly and in a memorable way. Paleontologist Richard Cowen later good-naturedly ribbed me for seeing the world through green-tinted glasses.

I published a paper titled "The Garden of Ediacara" in a new journal, *Palaios*, in early 1986 (figure 5.1).[25] The informal reactions of more senior scientists were interesting. One scientist found it unconvincing, arguing contentiously that he considered photosymbiosis to be a more "advanced" as opposed to primitive trait. He was apparently unaware of the extent to which living animals of "primitive" aspect are involved in a variety of photosymbioses.

Seilacher loved it. When we met at Bradley International Airport in Connecticut (he had accepted my invitation to be the Five College Distinguished Lecturer in Geology), he intimated that he wished that *he* had thought of the phrase.

Although the initial idea of photosymbiosis in the Ediacarans originated with Fischer, not me, its fuller development was mine, and I had given it an enduring name. My paper was cited by most of the scientists working on the Ediacaran biota. Fedonkin translated it as "Sad Ediakary" (Ediacaran garden) into Russian.[26] His commentary on the paper was as follows:

> An even earlier point of view is that the forms of the bodies of Ediacaran animals are optimized for internal photosymbiosis (Fischer, 1965). This idea has received wide support (Fedonkin, 1983; Seilacher, 1984; Conway Morris, 1985). It is possible that the availability of endosymbiotic algae made Precambrian Metazoa comparatively independent of external food sources, inasmuch as the multiplication in the bodies of animals of symbiotic algae could provide a significant part of the animals' food requirements, as is known to occur in several types of extant invertebrates. In fact, many Vendian organisms have leaflike forms, and the basic function of the leaves of plants is to capture light. Circumstantial evidence favors the existence of endosymbiotic, photosymbiotic algae in these Vendian forms, such as the fact that most fossils of the Vendian fauna have been discovered in shallow-water facies deposited in the marine photic zone. M. McMenamin conjectures that in some sense flat Precambrian organisms were ecological analogs to deep sea

The Garden of Ediacara

MARK A. S. McMENAMIN
Department of Geology and Geography
Mount Holyoke College
South Hadley, Massachusetts
01075-1484

PALAIOS, 1986, V. 1, p. 178–182

The faunal change across the Proterozoic-Cambrian transition can be interpreted as a shift from marine communities composed largely of soft-bodied,
possibly photoautotrophic animals to
communities dominated by heterotrophic
Metazoa. Convincing evidence for
macrophagous predation (damaged prey,
actual fossils of predators, and anti-
predatory prey morphologies) first appears
in the fossil record during the Lower

they do indeed represent animals in the sense of modern Metazoa; Seilacher, 1984) are referred to as the Ediacaran, Ediacarian, or Vendian soft-bodied fauna. Much attention has recently been focused on the phyletic significance of the Ediacaran fauna, especially the un-resolved conflict between Glaessner's (1984) notion that many members of this fauna belong to extant phyla and Seilacher's (1983, 1984) proposal that Ediacaran creatures have no phyloge-netic relationship to later Phanerozoic faunas. Perhaps the most interesting aspect of this debate is Runnegar's (1982) and Seilacher's (1984) contention that some members of the Ediacaran fauna lived as diffusion animals by be-coming flat and thin. Thin, ribbonlike

FIGURE 5.1: A portion of the first page of the original "Garden of Ediacara" article, published in 1986 in the journal *Palaios*.

pogonophorans called vestimentiferans (McMenamin, 1986). These pogonophorans, using endosymbiotic chemosymbiotic bac-teria as their biochemical energy source, have been called "auto-trophic animals" (Felbreck, 1981).

Calling the Vendian fauna photoautotrophic, M. McMenamin thinks that this unique biota ("The Garden of Ediacara") was adapted to existence in oligotrophic marine conditions, and at a later time was transformed into the heterotrophic Cambrian biota. A possible external stimulus for these changes was the Vendian-Cambrian phosphogenesis episode, as shown by the wide distrib-ution of phosphorites and the elevated content of P_2O_5 in sedi-ments of this interval (Cook and Shergold, 1984; Rozanov, 1985). Such phosphorites can form as a result of oceanographic phenom-ena that cause upwelling[27] on an unprecedented scale, which, in its own turn, must be leading to eutrophication of ocean basins, with corresponding changes in the feeding strategies of organisms: Growth of oceanic primary producers contributed to the spread of heterotrophs, including numerous predators.

The challenge of predation pressure and its accompanying high selective forces was answered by armor composed of skeletal elements and by secondary defense features such as the ability to escape predators by burrowing into sediment. In the Precambrian animals lived as autotrophs, with their endosymbionts functioning as the primary producers; that is, the trophic pyramid was so short that these animals were unlikely to develop skeletons or to restructure their essential feeding styles to gain the selective advantages of protection (McMenamin, 1986).

Detailed studies of Precambrian invertebrates have demonstrated the considerable importance of particular large groups that are presumably still extant. The unfamiliar or even strange body plans of these animals, representing part of the early radiation of multicellular organisms, is not encountered in more recent paleontological records. This renders the interpretation of these Precambrian fossils more difficult, but does not make their study less interesting or less important for our understanding of the historical course of the spread of ancient multicellular organisms.[28]

The Garden of Ediacara paper was cited by Donovan in a review article published in *Nature* on September 24, 1987.[29] All these things were well timed for me professionally, and shortly afterward I was awarded a prestigious Presidential Young Investigator award from the National Science Foundation. The award provided ample funding for 5 years.[30] I feel that the early success of Garden of Ediacara was at least partially responsible for the award.

Every success of this sort will inspire imitations and critics.[31] Guy M. Narbonne and J. D. Aitken, in their 1990 description of Ediacaran fossils from the Sekwi Brook region of northwestern Canada, state that, "[The] occurrence of a predominantly *in situ* benthic fauna in muddy slope deposits below storm wave-base is not consistent with the hypothesis that these taxa functioned exclusively as photoautotrophs."[32]

Narbonne and Aitken based their rejection of Ediacaran photoautotrophy on inferences of the depositional environment for the fossil-bearing Sheepbed and Blueflower formations. They interpret these strata as deep water, based primarily on an interpretation of some of the clastic layers similar to a type of deep water sediment known as a turbidite.

The paleobathymetry of this and indeed all putative deepwater Ediacaran fossil localities is a contentious topic. With regard to the Sheepbed Formation, Narbonne and Aitken argue that turbidites (deposits from turbidity currents) are present in the formation but then apparently reverse themselves and interpret the strata as slope deposits that *advanced* over supposed deep sea deposits of the lower Sheepbed. Narbonne and Aitken do not address the possibility that these supposed turbidites in the Sheepbed and Blueflower are tempestites (storm sand) deposits rather than deposits on an abyssal fan (turbidite).[33]

The depositional setting of the Sekwi Brook site was part of a newly formed rift ocean, with active block faulting and locally steep seafloor slopes in nearshore areas.[34] Water depths in these rift basins were, initially at least, very shallow. Also, this area, having just participated in the breakup of the Proterozoic supercontinent Rodinia, would have been a seismically active region in the late Proterozoic. Seismic activity has long been implicated as a trigger for events of lateral sedimentation.[35]

Based on the assumption that the Sekwi Brook biota lived in a "deepwater, basin slope setting below storm wave-base," Narbonne and Aitken imply that the organisms lived below the photic zone. Regarding this question, Hans J. Hofmann commented that he was "not prepared to say" whether the strata were deposited below the photic zone.[36] Crimes and Fedonkin noted that the Sekwi Brook biota occurred "not far below wave base."[37] Because tempestites probably occur in the sequence, the Sekwi Brook biota did not in fact live below storm wave base.[38]

Narbonne, apparently stung by criticism of his paleobathymetry assessment for the Sekwi, enlisted the support of sedimentologist R. W. Dalrymple to help with the research. In their joint paper on the sedimentology of the Sheepbed Formation, they conclude that the rocks were deposited in water depths of 1–1.5 km.[39] They base this argument on the interpretation of abundant ripples in some parts of the Sheepbed Formation as having been the result of cryptic deep marine currents that sweep along the edge of the continental shelf, leaving deposits called contourites. It is unclear whether there would have been sufficient ocean in the Australo-American Trough (the ancient ocean in which the Sheepbed Formation was deposited) to form contourites at this time, and the data set Dalrymple and Narbonne present in support of the contourite current direction is based only on a handful (14 measurements) of data.[40] On September 26, 1996, I asked Dr. Dalrymple by e-mail,

> I read with interest your recent paper with Guy Narbonne entitled "Continental Slope Sedimentation in the Sheepbed Formation." Is there any way to distinguish between current ripples formed in contourites and current ripples formed by shallow water processes,

other than a case made by inferring the sedimentologic context? I understand the argument from paleocurrent data, but your modal value in the rose diagram is based on only 10 data points. In other words, how confident are you that these are indeed deep water deposits? Many thanks for considering this question.

Dalrymple replied to my question the next day, with a thorough and thoughtful response. First, he asked whether contourite ripple markings are distinctive for deep water. To this he replied "no" because ripples just like these can form in any of a number of different types of sedimentary environments, most of them shallow water. Wave ripples, on the other hand, are distinctive of shallow water, and represent where wave energy has impinged on the bottom sediments. No wave ripples are seen at Sekwi Brook.

Second, he addressed the question of how he and Narbonne knew there were contourites present at Sekwi Brook. Here he seems to assert that they are contourites because they lack features typical of deepwater deposits (such as the graded bedding seen in turbidites), and because they look different from the other, surrounding strata.

Third, he responded to my request about his confidence in his and Narbonne's interpretation that these are deepwater deposits. The case he makes here is primarily a negative one, based on the absence of wave ripples and hummocky cross-stratification, and on the overall stratigraphic context. Hummocky cross-stratification, associated with storm deposition (and hence shallow water), occurs at Sekwi Brook, but Dalrymple and Narbonne dismiss it as chunks of sediment transported into the area. To me, the argument for deep water seems particularly weak here.

There is unquestioned evidence (sediments deformed by sliding) that the sediments were deposited on a slope, but this does not necessarily mean that the strata formed in deep water—only that there were locally steep slopes. Lastly, Dalrymple and Narbonne argue that the water depth at the site must have been 1–1.5 km because Sekwi Brook is 35 km offshore from the North American continental shelf. This argument assumes a steady increase in depth as one went offshore in the Australo-American trough, and may not be a safe assumption for a young rift ocean known to have been characterized by abundant block faulting,[41] with blocks tilted both away from and toward the continental shelf. Indeed, these blocks could be responsible for the locally steep slopes. I conclude, therefore, that the interpretation of water depth of deposition at Sekwi Brook is still very much an open question, but I appreciate Dr. Dalrymple's rapid and thoughtful response.

Can we get better constraint on the depths at which the Ediacaran creatures were deposited? And exactly how deep *is* the photic zone? The photic zone is defined in most textbooks as ending at 100 m water depth. However, photoautotrophs and photosymbiotic animals are found at marine water depths in excess of 100 m.

These questions can now be authoritatively addressed, thanks to the research of Dietrich Schlichter, a physiological ecologist at the Zoological Institute of the University of Cologne, Germany. In 1988, Schlichter and his co-authors found the fungiid coral *Leptoseris fragilis* engaging in photoautotrophy at depths of 100–145 m (color plate 13).[42] This coral lives in the Red Sea, a rift ocean that, tectonically and oceanographically speaking, is quite similar to the rift ocean that once formed the western (the Australo-American Trough) boundary of North America during the breakup of supercontinent Rodinia.

Large photosynthetic algae have been recorded at a depth of 268 m off the coast of San Salvador Island in the Bahamas.[43] This is considerably deeper than the estimate made above of storm wave base (100 m), indicating that the Sekwi Brook biota could easily still have lived within the photic zone *sensu latu*.[44]

In 1994 Schlichter sent me a large color print, taken from a submersible, of *Leptoseris fragilis* living at 120 m water depth in the Red Sea. These olive green corals, photosynthesizing at great depth (Schlichter calls it the Red Sea Twilight Zone), lent the photo a feeling of serene tranquillity (color plate 13). Only a tiny fraction of the sunlight incident on the surface of the ocean reaches these depths, most of it in the blue and green band of the solar spectrum, not particularly useful wavelengths for photosynthesis.

Schlichter found that *Leptoseris fragilis* engages in photosynthesis in an environment in which less than 1 percent of ambient sunlight reaches the coral. How are the corals able to photosynthesize at such great depth? They use several tricks to help their photosynthesizing symbionts and to enhance their access to photosynthetically active radiation. First, the symbionts are arrayed within each coral's tissue in a monolayer or staggered fashion, so that the symbionts avoid shading one another. Second, pigments found within the coral absorb the short wavelengths of light available at these great depths and reradiate them at more photosynthetically useful wavelengths. The implication of all this is that photosynthesis in animals can proceed at vanishingly low light levels. Other organisms do it even deeper; recall the record noted above of photosynthesis at a truly astounding 268 m depth.

The great geochemist Vladimir I. Vernadsky inferred an exceptional case of photosynthesis at depth, linking it to a downward extension of the photic zone due to bioluminescent bacteria:

> The phosphorescence of organisms (bioluminescence) is composed of wavelengths identical to those of solar radiation on the earth's surface. This secondary luminous radiation allows green plankton over an area of hundreds of square kilometers to produce their chemical work during times when solar energy does not reach them. It becomes more intense with depth.
>
> Is the phosphorescence of benthic organisms a new example of the same mechanisms? Is this mechanism causing life to revive several kilometers below the surface by transmitting solar energy to depths it could not otherwise reach? We do not know, but we must not forget that oceanographic expeditions have found green organisms living at depths far beyond the penetration of solar radiation. The ship *Valdivia* found, for example, *Halionella* algae living at a depth of about two kilometers.
>
> The transportation by living matter of luminous energy into new regions in the form of thermodynamically unstable chemical compounds and also of secondary luminous energy phosphorescence causes a slight provisional extension of the domain of photosynthesis.[45]

Photosynthesis in 2000 m of water seems implausible but Vernadsky gives cogent reasons why it might occur. If bioluminescent microbes create light with their luciferins (oxygen-protecting enzymes) bioluminescing thousands of meters below the surface, one can conjure an image of a Proterozoic seafloor illuminated at great depth. Even if Dalrymple and Narbonne are correct in their 1.5-km depth estimate for the Sheepbed Formation, photosynthesis *still* could be occurring at the bottom of the Australo-American Trough. One can imagine this Proterozoic sea, glowing eerily even at considerable depth with the light of bioluminescent plankton.

Startling new research results reinforce these enlightening possibilities; for example, biologically important light has been detected at the bottom of the sea. The modern deep-sea vents provide important insights into the nature of the ancient Garden of Ediacara.

A deepwater shrimp, *Rimicaris exoculata* ("eyeless shrimp of the fissure"), has been found in the Snake Pit hydrothermal vent field on the Mid-Atlantic Ridge, the place where new lava is added to the Atlantic seafloor as it spreads apart. *Rimicaris exoculata* indeed has no eyes, but each animal has an unusual pair of bright strips on the upper front part of its carapace. Close examination by biologists revealed this pair of strips to be weird organs of vision, away from the eye sockets, where the eyes had been evolutionarily lost. The organs of light detection on the shrimps' back are called hypertrophied dorsal eyes.

Until the discovery of *Rimicaris exoculata* in 1989, it was thought that all deep-sea vent animals were blind. In 1996 a second species (*Rimicaris* sp.) of vent shrimp was found, also with an enlarged dorsal eye. These unexpected organs are specialized for detecting extremely faint light. In the vent *Rimicaris* species, the faint light passes through a cornea underlain by a membrane carrying a massive array of photosensors.[46]

The vent shrimp use these unusual eyes to see the very faint light emitted by the vent itself.[47] This ability is useful to the shrimp because it allows them to maintain the proper distance from the food sources of the sulfide-rich vent waters. Too far away and the shrimp starve; too close, and shrimp cocktail.

This research has led to some fascinating speculation on the origins of photosynthesis. Could there be enough light emanating from the vents to power photosynthesis? Apparently there is.

The current world record holders for photosynthesis at faint levels of light are photosynthetic bacteria living approximately 80 m below the surface of the Black Sea. Here the bacteria receive a trillion photons of light per square inch per second, as compared to a tree on a sunny day, which receives a quintillion photons per square inch per second. The tree receives light ten thousand times more intense than that captured by the Black Sea bacteria.[48]

Ever since her discovery of the deep vent light, Cindy Lee Van Dover had considered the possibility that the vent light was sufficient for some type of photosynthesis. Colleagues had told her that this was a stupid idea, but it now appears they were wrong. The vent light is of approximately the same intensity as the light used by the Black Sea bacteria. Thus, photosynthesis is possible at the seafloor, using nonsolar light.

The intriguing possibility exists that photosynthesis *originated* at deep sea vents. The most ancient types of bacterial chlorophyll are most sensitive to exactly the frequency bands measured by Van Dover at the vents.[49] It would be highly ironic if the use of bright sunlight by photosynthesizers was an afterthought, developed long after the origin of photosynthesis with faint vent light in the deep sea.

A terrestrial analog to photosynthesis at low light levels is algal or cyanobacterial communities living under or within rocks. Photosynthesizers have been reported occurring at up to 61 mm below the soil surface and under quartz rocks of 109 mm in thickness.[50]

My original Garden of Ediacara concept, following both the Fischer inference of photosymbiosis and Seilacher's[51] inference that some of the Ediacaran forms might have been chemosymbiotic, admitted both photosymbiosis and chemosymbiosis as feeding strategies in the Garden. To these two I added a third, direct absorption of dissolved nutrients, or osmotrophy.[52] Aquatic animals and other organisms can feed in this way, in what amounts to a natural version of hydroponics. Of course, there is nothing to stop organisms from combining feeding strategies, as many living organisms do in what is called mixotrophy.

The chemosymbiosis possibility gained plausibility by the presence of the modern deep-sea vent communities, with a food chain driven by hydrogen sulfide and chemosymbiont bacteria in the tissues of the rather large animals at great depth. As alluded to by Fedonkin in his discussion of the Garden of Ediacara, Felbreck called giant tube worms of the vent faunas "autotrophic animals."[53] The larvae of these tube worms have great abilities of dispersal, allowing them to colonize the widely separated and short-lived hydrothermal vents on the seafloor.[54] The most important characteristic of these modern communities is that they lack the ordinary dependence on sunlight to energize the food chain. Chemosymbiosis is widespread even away from deep ocean vents and has been recognized in 30-million-year-old fossil sponges associated with marine cold seeps.[55]

So if some Ediacaran forms were chemosymbiotic, they could have thrived in the absence of light. This possibility led Richard A. Fortey to question whether the Garden of Ediacara theory could be critically tested: "But some are found in deep water sediments; never mind! these can be supported by a sulfur-based metabolism!"[56]

Fortey mistakenly implies here that Garden of Ediacara theory relies chiefly on its photosymbiosis aspect. I expanded on the osmotrophy aspect of Garden of Ediacara theory in a paper published in the journal *Invertebrate Reproduction and Development*.[57] In this paper I argued that osmotrophy may have been the most important of the three (photosymbiosis, chemosymbiosis, and osmotrophy) feeding styles in the Proterozoic.

Vernadsky pointed out the importance for the biosphere of the difficult-to-detect and "usually neglected" trace amounts of organic compounds occurring in natural waters. He estimated the mass of this organic matter present in natural waters to be at least several quadrillion tons. The material is derived from heterotrophic organisms as well as autotrophic organisms. Many of the compounds are nitrogen rich[58]:

> Nature continually makes these materials visible without recourse to chemical analysis. They form fresh water and salt water foams, and the iridescent films which completely cover aquatic surfaces thousands and millions of square kilometers in area. They color rivers, marshy lakes, tundras, and the black and brown rivers of tropical and sub-tropical regions. No organism is isolated from these organic compounds, even in the solid earth itself, because they are continually penetrating it by means of rain, dew and solutions of the soil.
>
> The quantity of dissolved and colloidal organic matter in natural water varies between 10^{-6} and 10^{-2} percent, and has a gross mass amounting to 10^{18} to 10^{20} tons, *mostly in the ocean* [italics mine]. This quantity seems to be greater than that of living matter. The idea of its importance is slowly entering contemporary scientific thinking. Even with the old naturalists we often come across an interpretation of this impressive phenomenon, sometimes from an unexpected point of view.

Using a new technique for the oxidation of organic compounds in seawater, Sugimura and Suzuki reported a quantity of dissolved carbohydrates and proteins in the world's oceans that is at least twice as large as previous estimates of the amount of dissolved marine organic carbon.[59] In other words, there are unclaimed nutrients in seawater, free for the taking.

These measurements may explain anomalous observations made in the 1960s by Gordon Riley of Yale University.[60] Riley reported the appearance of particulate material in seawater in the winter and the early spring. According to Dr. J. R. Vallentyne of Cornell University, commenting on Riley's results:

> The transformation from dissolved to particulate is obviously not biological since it can happen by simply bubbling air or nitrogen through sea water that has been filtered to remove all particulate material. . . . [Appearances of these organic particulates from solution] usually follow plankton bursts and they do take nitrogen out of the water.[61]

Absorption of dissolved nutrients from seawater has been demonstrated in all living classes of echinoderms and may provide an important source of food in addition to particulate matter obtained by filter feeding and other methods.[62]

The Ediacarans aside for the moment, the Proterozoic world ocean had many large algae (known from carbonaceous megafossils) that used glycolate dehydrogenase to oxidize glycolate to glyoxylate during photorespiration. Glycolate dehydrogenase similar to that found in chlorophyte algae is found in the prasinophytes and enigmatic cryptophytes. Under high atmospheric oxygen tensions, a likely condition during the late Proterozoic,[63] users of glycolate dehydrogenase[64] would have been compelled to excrete larger amounts of unmetabolized glycolate into seawater than under lower oxygen regimes.

I hypothesized in 1993 that any abundance of unclaimed marine glycolate could have nourished large Proterozoic osmotrophs, particularly forms with a large surface area.[65] Supplemental nutrition by phototrophy via microbial symbionts is also likely if these organisms were translucent and resided in the photic zone.

In a remarkable 1991 paper that addressed natural hydroponics and the issue of dissolved nutrients in seawater, Schlichter and G. Liebezeit reported that the symbiont-bearing soft-bodied coral *Heteroxenia fuscescens* released dissolved free amino acids into the water column.[66] In other words, the photosynthetic output of the coral's photosymbionts exceeds the demand of *both* the symbionts and the host. Other carbon compounds are also released by the coral. Schlichter and Liebezeit caution that in calculating the amount of organic carbon transferred from symbionts to host, it is important to note that surplus nutrients are not necessarily used by the host.

This leads me to ponder an interesting potential feedback system in the Garden of Ediacara. Large Proterozoic host organisms and their symbionts generated more dissolved organic compounds, including amino acids, than they could use. As a result, they released excess nutrients into the water. Oxygen levels were high, higher than today, so any users of glycolate dehydrogenase release unmetabolized glycolate, further adding to the hydroponic nutrient broth. As in the story of the stone soup, each member of the biota adds its own contribution to the marine mixture until it becomes a rich broth.

This could lead to a self-perpetuating biospheric exchange of dissolved *organic*[67] nutrient abundance and perhaps might explain why large predators were apparently absent during the Proterozoic. To put it in somewhat anthropomorphic terms, why bother evolving into a large predator if the marine medium consists of a nutrient-rich broth, a free lunch, that can be conveniently used at any time for food? Why bother eating anyone else?

At least in the photic zone, the more organisms that partake of the broth, the more are released back into the water as leaked photosynthate. Certainly the constraints of limitation of certain key nutrients (particularly phosphorus) would still apply, but with so much dissolved organic carbon and other nutrients in seawater, being distributed throughout Mirovia[68] by ocean currents, one can image how an entire marine biota might be sustained in this way. The system would be self-stabilizing because as new photosymbionts were added to the marine biosphere, they could create nutrient excesses that could be released back into the water. Expansion would occur until the limits of some vital mineral nutrient were reached. This biotically beneficial situation might extend not only to the bottom of the sea, but also into marine sediments.

Seilacher was first to note that certain Ediacaran forms such as *Pteridinium* might have lived partly or entirely buried within the sediment. This view is still quite controversial, with some paleontologists arguing that *Pteridinium* grew flat on the seafloor and was fossilized by being covered by storm sands. Seilacher has made sketches of *Ernietta* and *Pteridinium* partly submerged or entirely submerged, respectively, in seafloor sediment. This obviously could pose a problem for the photosymbiosis part of Garden of Ediacaran theory, for it is generally felt that photosynthesizers, or at least their photoactive parts, live above the sediment surface. But is this really always the case?

Here are notes from my research notebook (p. 42) of February 9, 1994:

I read Schlichter's article, just in from interlibrary [loan]. It is good for Garden of Ediacara hypothesis—*Leptoseris* is a high efficiency phototroph coral able to live at 145 m. Zooxanthellae[69] are stacked to avoid shading each other, and the animal cells have granules which contain autofluorescent pigments transforming short wavelengths into longer ones. . . .

My thoughts turn to the Namibian *Pteridiniums*. Surely these forms seem to be trying to maximize surface area in a sense parallel to the earth's surface. Perhaps if they are partly buried in clean sand, they still get enough light [figure 5.2]. Light basin for capture of sunlight. If Antarctic endoliths [rock-dwelling lichens] do it, why not *Pteridinium* in the sea?

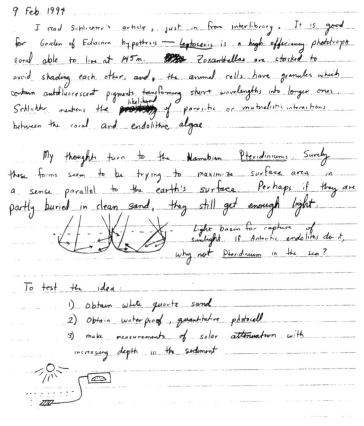

FIGURE 5.2: Sketch from my research notebook of February 9, 1994.

To test the idea:

1. Obtain white quartz sand.
2. Obtain waterproof, quantitative photocell.
3. Make measurements of solar attenuation with increasing depth in the sediment [figure 5.2].

Mount Holyoke student Lori Bennett and I set out to examine this idea. For a class project, Bennett built, with painstaking technical finesse (and a lot of black electrician's tape), a waterproof dark chamber with glass bottom. Under the glass, within the same dark chamber, was a light meter we borrowed from the physics department. Using this experimental chamber, Lori and I tested the transmission of light through layers of varying thickness of different types of both wet (fresh and salt water) and dry sand.

We reported our results as a poster session at the Geological Society of America annual meeting in Seattle.[70] It might seem to be counterintuitive at first, but light actually transmits quite a bit better through wet sand than it does through dry sand. We found that photosynthetically useful light from both solar and artificial light sources transmits through a thickness of up to 23 mm of salt-water-saturated coarse sand. We were using dark and impure sand from the coast of Maine, not a selected pure white quartz sand. We used these results to suggest that it is likely that sand-loving organisms would be able to photosynthesize beneath the sediment.

We went on to say that the "soft" members of the Ediacaran biota had a head start toward fossil preservation, first because they were already buried while alive, and second because if they were living in sand they must have had a sand-resistant cuticle. We noted that fossils of the Namibian slab excavated by Seilacher's team (color plate 12) have a preserved relief of generally less than 25 mm-the living depth of *Pteridinium* in Namibia may have been set by light availability, which Bennett and I demonstrated diminishes to very low levels at 23 mm. Finally, we noted that the photic zone must now be extended into the shallow geosphere (which starts at the surface of the sediment) in order to encompass the zone of light penetration within subaqueous sediments, and that *Pteridinium's* shape was suggestive of photosynthesis within the sand.

The poster generated positive comments from several senior paleontologists at the meeting. After the meeting, however, I learned from anonymous reviewers that to a certain extent we had reinvented the wheel.

In 1937, as the clouds of war were gathering over Europe, a German scientist at the University of Kiel named Eric Schulz published a paper introducing the term *farbstreifensandwatt*.[71] This very Germanic term translates literally as "color-striped sand layers."

Schulz had studied an unusual community of photosynthetic beach bacteria. These bacteria lived below the surface of the damp beach sand, and were alternatively inundated and exposed by the tides. Cutting into sands of North Sea beaches does indeed reveal color-striped layers, greens and reds a short distance below the surface. Each colored layer represents a particular bacterial community.

In 1940 and 1941, near the high-water mark of Nazi Germany, German scientist Curt Hoffmann ran experiments designed to reveal more information about the *farbstreifensandwatt*. He was not ready to publish until 1948, at which point he had to proceed in the ruins of postwar Germany.[72] Hoffmann complains in the paper about war-related losses of equipment and notes.

Hoffmann's experiments measured the transmittance of light though sand layers of varied thickness. He varied the water content of the sand (grain size 0.2 mm) from dry to fully saturated. He used a 200-watt artificial light source; light was passed through the sand and through 9 cm × 12 cm glass plates. Thin glass strips were abutted to the edges of the glass plates with cured Canada balsam as glue, forming a glass-sand-glass sandwich.[73] Light was then passed through the sand and the glass in a dark chamber to a light meter[74] 3.5 cm in diameter.

Hoffmann placed colored filters between the light source and the sand to test the transmissivity of light of various wavelengths. He found that red light penetrated the deepest, green light was next, and blue had the shallowest penetration. Light transmits considerably further through damp sand (17.55 percent water) than through dry sand (2.89 percent water). Hoffmann's explanation for this, which I endorse, is that water saturation reduces the amount of loss due to reflectance of light from the surfaces of sand grains. This is why wet sand appears to be darker than dry sand when viewed from above; the light is absorbed into the sand rather than being bounced back.

Hoffmann tested the transmittance through sand layers only up to 6 mm in thickness, but it is clear from his table 1 (p. 51) that transmittance could have gone considerably deeper, particularly with the red light passing through wet sand. However, Hoffmann was primarily interested in the light reaching the shallow *farbstreifensandwatt*. A color photo of the *farbstreifensandwatt* can be found in the book *Biostabilization of Sediments.*[75]

In his results section, Hoffmann quite reasonably concluded that the reason for the stripes in the *farbstreifensandwatt* is that the bacteria in question are specialized for capture of particular parts of the spectra. In other words, the lowest *farbstreifensandwatt* layer is composed of bacteria specialized to photosynthesize with the deeply penetrating red light. I wonder whether, as is the case with the symbionts of *Leptoseris fragilis* coral, these deeper bacteria also have autofluorescing pigments or other tricks to maximize use of faint light.

It seems quite likely, if indeed it was buried in life, that *Pteridinium* used such tricks. The more horizontal vanes of the form are turned upward, like the collectors on a solar panel (figure 5.2). The vanes themselves might have acted as a type of bio-fiber optics, light pipes that picked up light with their edges at the sediment surface and transmitted light downward even further than 25 mm. Viewed in this way, *Pteridinium* can be considered an ancient paradox, a subsurface light-harvesting organism.

Modern organisms may do this as well. I have eaten carrots with photosynthetic streaks running several centimeters down the core of the carrot. Light was apparently transmitted within the carrot, causing the ordinarily orange flesh of the root to turn green. The hollow white hairs of polar bears transmit light (and hence warmth) to the animal's black skin.

Supposed fossil symbionts have been reported in *Pteridinium*'s neighbor fossil *Petalostroma kuibis*, but these bodies may be artifacts of preservation rather than actual fossils of symbionts.[76] I have found what I interpret as direct evidence of photosymbiotic activity in *Pteridinium*.

The *Pteridinium* specimen Bruce Runnegar had asked me for in Namibia (chapter 4) proved to be the critical object in this piece of research. The specimen, you will recall, consisted of the upright vane of the *Pteridinium* plus one of the "bathtub" horizontal vanes; the other had been lost and was not even present on the original boulder from which my hand sample was taken. Back in my laboratory in Massachusetts, I decided to X-ray this sample to see whether hard X rays revealed any interior structures of interest. The rock showed faint cross-bedding in the layers immediately above the horizontal vane, and headed toward but not touching the vertical vane (figure 5.3), so I surmised that perhaps the relationship between the cross-bedded layers within the rock and the vertical vane would tell me more about whether *Pteridinium* was indeed buried in life.

I was thinking to myself as I started the X-ray setup that this could be an involved process, trying to find an appropriate exposure time. I had no idea about how much time to set the exposure for, so I asked our technician Patricia Weaver to bracket a series of exposures at 3, 5, 7, and 10 minutes. I wanted to direct the X-ray beam right down the axis of the *Pteridinium*, so I needed a holder to keep the obelisk-shaped specimen upright. I quickly improvised with a black plastic film canister with its lid off. The rock balanced neatly on end in the film can, and the first round of exposures were, to my relief, made without it falling off once.

On examining the negatives on my light table, I was quite pleased to see that many of the negatives appeared to be usable.[77] It turned out that my guess of an exposure time was exactly right. A few minutes in the darkroom yielded perfectly exposed prints (figure 5.4). The cross-beds were faintly visible, but the prints also held a surprise. On the prints, in the rock zone immediately above the horizontal vane, was a horizontal series of dark spots (bright spots on the negatives). These represented some type of electron-dense objects within the matrix of the specimen but apparently not in direct association with the fossil itself (figure 5.5).

What could these objects represent? I double-checked the other neg-

FIGURE 5.3: A Namibian specimen of *Pteridinium* showing an upright, exposed chaperone wall and one of the two "bathtubs." Length of fossil in view 6.5 cm.

atives, some with less than ideal exposures, and all of the ones that were fairly well exposed showed the same dots. Some of the negatives were taken at slightly different angles, so I was able to use their prints to create makeshift, imperfect (the angle was wrong) stereo pairs. The pseudo-stereo pairs seemed to indicate that the dots were in a plane parallel to the axis of the obelisk, but not all in a line. By analogy, consider the relative orientations of the true positions of the stars in Orion's belt. This meant that the dots were occurring in a horizontal zone a short distance above the horizontal vane.

What could this mean? My first thought was that the electron-dense objects were pyrite crystals, and indeed this is still my preferred interpretation. Recall that I had found large pyrite crystals in association with the base of *Ernietta* fossils (figure 5.6) in Dr. Pflug's home in Lich, Germany. I am loath to cut up the specimen because I would at some point like to make true stereo pairs and time has not yet permitted me to do this.

Let us assume for the moment that I am correct about the dots being pyrite. Pyrite forms in sediments when there is enough organic carbon to induce bacteriogenic reduction of sulfate.[78] The dots in my radiograph are forming *not in contact with the actual fossil remains*. The impli-

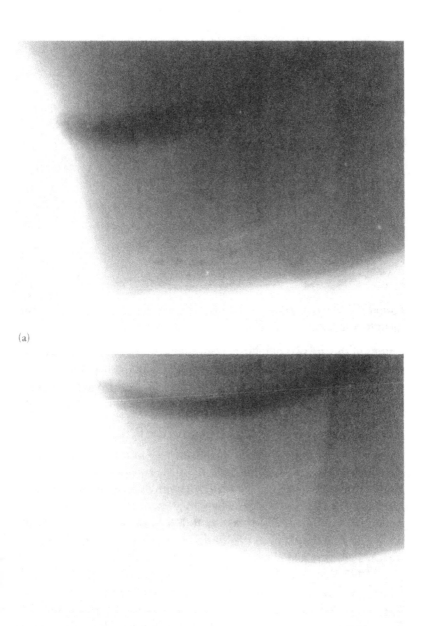

(a)

(b)

FIGURE 5.4: X-ray view along the long axis of the specimen portrayed in figure 5.3. Note electron-dense bodies, light layer (bedding horizon in sandstone), and dark curving arc (X-ray shadow from film can on which the specimen was balanced). (a) and (b) show the view at slightly different angles. Width of view approximately 3 cm.

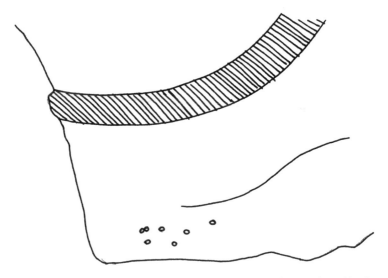

FIGURE 5.5: Interpretive sketch of the radiographs shown in figure 5.4. Width of view approximately 3 cm.

cations of this are twofold. First, *Pteridinium* was not completely buried in life.[79] There were only a few millimeters of sand (perhaps Hoffmann's 6 mm) filling the curving bottom of the bathtub, perhaps as passive ballast. The chaperone wall was largely exposed, as were the upright walls of the bathtub.

Second, these organisms were photosynthesizers. Like the part of the organism in the water column, the buried part of the horizontal vanes were releasing excess photosynthate into their environmental surround. Within the sand, these organic carbon compounds traveled some distance from the *Pteridinium*'s cuticle, where they encouraged sulfur-reducing bacteria to form pyrite.[80] The pyrites thus formed are the small dots. If the pyrite had simply been due to decomposition, it would have also occurred adjacent and parallel to the chaperone wall.

In 1995 Kenneth M. Schopf and Tomaz K. Baumiller of Harvard University presented an abstract with the amusing title "A Biomechanical Approach to Ediacaran Hypotheses: Weeding the Garden of Ediacara."[81] They argue in this abstract that flat Ediacaran forms such as *Dickinsonia* and *Phyllozoon* would have been unstable in the sense that they could have been easily picked up and transported by marine currents. Their arguments, based on measurements of the critical slip velocity of models of the flat Ediacaran genera, are correct and are decidedly applicable to a

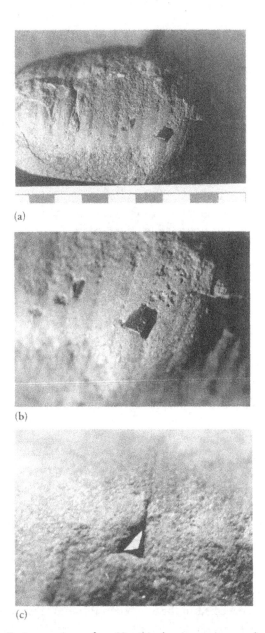

(a)

(b)

(c)

Figure 5.6: *Ernietta* specimens from Namibia showing pyrite crystals at the base of the specimen. (a) Large cubic pyrite crystal visible at base of fossil. Scale bar in centimeters. (b) Enlarged view of specimen in (a). (c) Corner of a pyrite cube peeking out of the base of a second specimen of *Ernietta.*

form such as *Pteridinium* with upturned edges. Their results do not indicate that no flat Ediacarans could have lived at the sediment surface—only that in order to do so they would have had to adhere to the sediment surface in some manner or have been weighed down with internal sand. Recall the mucous-holdfast hypothesis for *Vermiforma*.

Accepting the Garden of Ediacara as a viable working hypothesis, we can move forward with new research questions. The first thing I would like to see in this regard are *in situ* occurrences of Ediacaran fossils with inclinations indicating whether they had, in life, a preferred orientation to maximize light capture. The *Cyclomedusa* specimen from Mexico (chapter 9) may be an example of this, but this possible example is isolated and needs to be bolstered by more specimens. If such a find were to be made, it would be tremendously useful, for in addition to providing additional support for the Garden of Ediacara hypothesis, it would provide a valuable cross-check for paleomagnetically determined values of the paleolatitude at which the rocks bearing the fossils in question were deposited. Everything we actually know in the historical sciences comes from cross-checks of this sort.

It appears that photoautotrophy, osmotrophy, chemoautotrophy perhaps, and mixotrophic combinations of these three were the dominant feeding styles for large marine creatures for the duration of the Garden of Ediacara. The Garden of Ediacara hypothesis may be tested by, say, attempting to locate a *Dickinsonia* with a predatorial bite taken out of it. A recent paper makes a playful jab at the Garden of Ediacaran theory, noting that the main localities in the Namibian desert, the outback of Australia, and the Winter Coast of northwestern Russia "might seem unpromising Edens."[82] But in their main illustration the frondose Ediacarans are painted green.

Challenges to the Garden of Ediacara hypothesis have already begun. An article recently published in *Systematic Biology*[83] argues that the Proterozoic body fossil *Bomakellia* is closely related to the anomalocarids, a voracious group of very large Cambrian predators.[84] The author goes so far as to identify what he interprets as eyes on the only known specimen of *Bomakellia*, collected in the White Sea region of Russia (figure 5.7). Although this interpretation is less than compelling, the fossil could nevertheless represent an animal with eyes.[85] Whether it is related to the Cambrian anomalocarids, as the author argues, is even less certain, and whether *Bomakellia* is a predator is less certain still, but remains a possibility, especially in this very late Ediacaran horizon. Consider figure 2.16, the erniettid from Nevada, another very late Ediacaran form. This

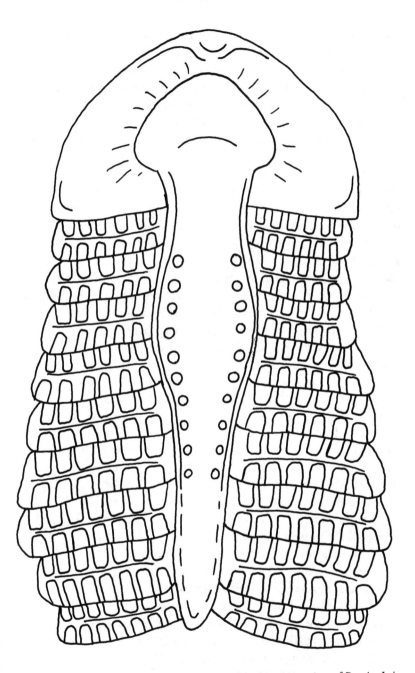

FIGURE 5.7: *Bomakellia*, an Ediacaran from the White Sea region of Russia. It is claimed to have been an arthropod-like form, but its relationship to true arthropods has yet to be convincingly demonstrated. Length of specimen 8 cm.

specimen has what might be an arcuate slice out of the end of it, and it also appears that the tubes have regrown. Could this be predatorial damage to a very late Ediacaran form by the earliest large predator?

Some of my colleagues have wondered publicly about the idyllic connotation of the phrase *Garden of Ediacara*. In a sense it is idyllic, but only on a macroscopic level. I am not making moral judgments about the Garden of Ediacara in comparison to our modern heterotrophic system. It would be ludicrous to attempt to do so. If I am right about this episode in earth history, however, everyone will have to very carefully rethink the implications of my discovery of this extinct ecosystem. In the Garden of Ediacara, the medium (sunlight, seawater, and its dissolved contents) and the message (free lunch) were the same. But nothing in life is truly free, and even this free lunch came with a price. Under conditions of nutrient and sun bath, might not prospects appear dim for the evolution of brains and, eventually, intelligent life? We return to this thought in chapter 12.

I don't think that there is anything wrong with considering the implications of the Garden of Ediacara in a wider sense. Are there fruitful analogies here for the conduct of, say, foreign policy? Do socialism and capitalism have, fundamentally, an ecological basis? I don't know the answers to these questions, but it might be enjoyable to speculate on them. On a more serious note, one that has become more serious with discussion of the possibility of Martian fossils, it has been argued that the fate of our species is to transport life to other planets, to become the agents of a directed panspermia. If so (and I can think of worse fates), we will require a profound understanding of the types of ecosystems (Garden of Ediacara, heterotrophic ecosystems, Hypersea) permissible in a bewildering variety of extraterrestrial settings. We will also have to understand how one ecosystem might transform into another. Understanding Garden of Ediacara theory, and applying its corollaries, could help plot a course for the future of the human species. For if we colonize space, we, like the Ediacarans, will face a constant fight against the threat of starvation and against low concentrations of food and energy.

As a final illustration of the Garden of Ediacara, consider these additional species from Namibia. Namibia has produced a variety of frond fossils that do not belong to the genus *Pteridinium*. When I first learned of the frond fossil *Nasepia altae* in the early 1980s, I noted its similarities to *Marywadea*, *Spriggina*, and *Dickinsonia*.[86] Another supposed sprigginid had also been described from Namibia.[87]

Nasepia is known from distorted and damaged (probably by a storm event) fossils, but as far as can be ascertained it seems to have consisted of approximately three petaloids joined together along a common axis. A better preserved fossil, named *Swartpuntia germsi*, has been described from the Nama.[88] *Swartpuntia* often has more than three vanes, and was perhaps a six-vaned version of *Nasepia*. A *Swartpuntia*-like form has recently been reported from the Wood Canyon Formation of the Spring Range, Nevada, a find which buttresses those reconstructions of the supercontinent Rodinia (see Chapter 8) placing Namibia close to the Mexican end of North America.[89]

Remarkable similarities exist between *Swartpuntia* and *Dickinsonia*. Reconstructed as standing upright on a stalk embedded in the sediment, *Swartpuntia germsi* had up to six petaloids arrayed around the axis of the stalk. The individual organisms have been preserved by flattening to form attractive butterfly-shaped fossils. Like *Pteridinium*, *Swartpuntia* specimens often occur in clusters. With petaloids arrayed in fan fashion to maximize capture of sunlight, an ancient stand of Namibian *Swartpuntia* or *Nasepia* would resemble a miniature Forest of Ediacara.

Notes

1. V. I. Vernadsky, *The Biosphere*. Translated by D. Langmuir, revised by M. McMenamin (New York: Nevraumont/Copernicus, 1997).

2. See p. 55 in S. Brand, ed., *The Next Whole Earth Catalog: Access to Tools* (New York: Random House, 1981).

3. J. P. McMenamin, "Observations on the Stomatal Structure of *Ilex opaca*," *Proceedings of the Indiana Academy of Sciences* 52 (1943):58–61.

4. A. Seilacher, "Late Precambrian and Early Cambrian Metazoa: Preservational or Real Extinctions?" in H. D. Holland and A. F. Trendall, eds., *Patterns of Change in Earth Evolution*, pp. 159–168 (Berlin: Springer-Verlag, 1984).

5. See p. 1211 in A. G. Fischer, "Fossils, Early Life, and Atmospheric History," *Proceedings of the National Academy of Sciences (USA)* 53 (1965):1205–1213.

6. Eumetazoans are the most familiar types of animals.

7. Page 8 in J. W. Schopf, B. N. Haugh, R. E. Molnar, and D. F. Satterthwait, "On the Development of Metaphytes and Metazoans," *Journal of Paleontology* 47 (1973):1–9.

8. See pp. 482–483 in M. F. Glaessner, "Pre-Cambrian Fossils," *Biological Reviews of the Cambridge Philosophical Society* 37 (1962):467–494.

9. See p. 485 in M. F. Glaessner, "Pre-Cambrian Fossils," *Biological Reviews of the Cambridge Philosophical Society* 37 (1962):467–494.

10. A trait he shared with R. Buckminster Fuller, who cautioned against showing unfinished work.

11. On December 1, 1982, regarding my paleobiogeographic theory for the Ediacarans.

12. M. A. McMenamin, "Evidence for Two Late Precambrian Faunal Province Loci," *Geological Society of America Abstracts with Program* 13, no. 7 (1981):508.

13. B. P. Radhakrishna, ed., *The World of Martin F. Glaessner* (Bangalore: Geological Society of India Memoir No. 20, 1991).

14. See p. 13 in J. W. Hedgpeth, "Evolution of Community Structure," in J. Imbrie and N. Newell, eds., *Approaches to Paleoecology*, pp. 11–18 (New York: Wiley, 1964).

15. I do not agree with Fischer, however, that the animals' microbial photosymbionts were needed primarily to produce oxygen.

16. Hedgpeth, 1964, p. 18.

17. Page 441 in A. C. Hardy, "On the Origin of the Metazoa," *Quarterly Journal of Microscopical Science* 94 (1953):441–443.

18. T. D. Ford, "Precambrian Fossils from Charnwood Forest," *Proceedings of the Yorkshire Geological Society* 31 (1958):211–217.

19. Personal written communication, 1988.

20. M. A. Fedonkin, "Ekologia dokembrijskikh metazoa belomorskoj bioty," in L. A. Nevessaya, ed., *Problemy ekologii fauny i flory drevnikh bassejnov*, pp. 25–33 (Moscow: Akademiya Nauk SSSR, Trudy Paleontologicheskogo Instituta, Tom 194, Izdatel'stvo "Nauka," 1983).

21. Translation by M. McMenamin, September 30, 1996.

22. In a recent paper Fedonkin cites Vernadsky but fails to cite any of the great symbiogeneticists (M. A. Fedonkin, "The Precambrian Fossil Record: New Insight of Life," *Memorie della Società Italiana di Scienze Naturali e del Museo Civico di Storia Naturale di Milano* 28 [1996]:41–48).

23. M. A. S. McMenamin, "Autotrophic Animals, Predators and the Precambrian-Cambrian Transition," *Geological Society of America Abstracts Plus Program* 11 (1985):475.

24. P. Hallock, "Algal Symbiosis: A Mathematical Analysis," *Marine Biology* 62 (1981):249–255; P. Hallock, "Drowned Reefs and Carbonate Platforms: Reef-Community Disruption by Nutrients Provides a Clue to the Paradox," *Geological Society of America Abstracts Plus Program* 17 (1985):602.

25. M. A. S. McMenamin, "The Garden of Ediacara," *Palaios* 1 (1986):178–182.

26. See p. 119 in M. A. Fedonkin, *Besskeletnaya fauna venda i ee mesto v evolyutsii Metazoa* (Moscow: Akademiya Nauk SSSR, Trudy Paleontologicheskogo Instituta, Tom 226, Izdatel'stvo "Nauka," 1987).

27. I love the Russian cognate (surely a loan word) for this word, *apvelling.*

28. Translation by M. McMenamin, September 12, 1996.

29. S. K. Donovan, "Mass Extinctions: Confusion at the Boundary," *Nature* 329 (1987):288.

30. As a bonus I received in the mail an award certificate signed by then-president Ronald Reagan.

31. D. Pendick, "The Ediacaran Garden," *Earth* 4 (1995):10.

32. G. M. Narbonne and J. D. Aitken, "Ediacaran Fossils from the Sekwi Brook Area, Mackenzie Mountains, Northwestern Canada," *Palaeontology* 33 (1990):945–980.

33. Narbonne and Aitken (1990) claim to be able to recognize Bouma Ta, Tb, Tbc, and Tc in the in the upper fossiliferous part of the Sheepbed and Bouma Ta, Tab, Tb, Tc, and Tbc in the lower three-quarters of the Blueflower Formation. Note that no more than two of the Bouma elements are juxtaposed in the supposed Bouma sequences cited above. Storm deposits can have a component of turbidity flow deposition and are not necessarily indicative of a subphotic zone depositional environment. The strong possibility that these clastic layers are reworked tempestites casts grave doubts on both the turbidite interpretation of these Canadian strata and on Narbonne and Aitken's paleobathymetric interpretations that follow. Evidence of lateral sedimentation (including, interestingly, an allochthonous raft of stromatolitic dolomite) is abundant in the Canadian strata but is not necessarily indicative of deepwater deposition.

34. G. M. Young, "The Later Proterozoic Tindir Group, East-Central Alaska: Evolution of a Continental Margin," *Geological Society of America Bulletin* 93 (1982):759–783.

35. Lateral sedimentation is the horizontal transport of loose sediment, as in above-water or submarine landslides.

36. Personal communication, March 30, 1995.

37. See p. 75 of T. P. Crimes and M. A. Fedonkin, "Evolution and Dispersal of Deep-Sea Traces," *Palaios* 9 (1994):74–83; T. P. Crimes and M. A. Fedonkin, "Biotic Changes in Platform Communities Across the Precambrian-Phanerozoic Boundary," *Rivista Italiana di Paleontologia e Stratigrafia* 102 (1996):317–332.

38. The term *storm wave base* has varied bathymetric connotations as currently used in the geological literature. Indeed, the water depth this boundary represents can vary over time in given sections (T. P. Burchette and V. P. Wright, "Carbonate Ramp Depositional Systems," *Sedimentary Geology* 79 [1992]:3–57). As used in recent publications, the depth of storm wave base has been interpreted as 60 m (C. E. Brett, A. J. Boucot, and B. Jones, "Absolute Depths of Silurian Benthic Assemblages," *Lethaia* 26 [1993]:25–40), 200 m, and possibly more than 200 m (W. J. Fritz and M. F. Howells, "A Shallow Marine Volcaniclastic Facies Model: An Example from Sedimentary Rocks Bounding the Subaqueously Welded Ordovician Garth Tuff, North Wales, UK," *Sedimentary Geology* 74 [1991]:217–240). Based on the depth at which *Thioploca* mats on the seafloor break up during the austral winter off the modern coasts of Peru and Chile, storm wave base apparently occurs at 60 m water depth (H. Fossing, V. A. Gallardo, B. B. Jørgensen, M. Hüttel, L. P. Nielsen, H. Schulz, D. E. Canfield, S. Forster, R. N. Glud, J. K. Gundersen, J. Küver, N. B. Ramsing, A. Teske, B. Thamdrup, and O. Ulloa, "Concentration and Transport of Nitrate by the Mat-Forming Sulfur Bacterium *Thioploca*," *Nature* 374 (1995):713–715).

The depth of storm wave base is a function of the wavelength of long-period swells associated with storms, and this wavelength in turn is a function of available fetch and the length of time storms spend over the water. Considering that the rift oceans formed by the breakup of Rodinia began as narrow seaways, the available effective fetch may have been very limited in comparison with modern oceans such as the Pacific or Atlantic. The storm wave base off the coast of western Canada at the time of the deposition of the Sekwi Brook faunas may well have been limited to 100 m water depth or less.

See also D. Palmer, "Ediacarans in Deep Water," *Nature* 379 (1996):114.

39. R. W. Dalrymple and G. M. Narbonne, "Continental Slope Sedimentation in the Sheepbed Formation (Neoproterozoic, Windermere Supergroup), Mackenzie Mountains, N. W. T.," *Canadian Journal of Earth Science* 33 (1996):848–862.

40. Also, the current direction determined is in any case rather close to directions in current structures measured in rocks not considered by them to be contourites.

41. G. M. Young, "The Later Proterozoic Tindir Group, East-Central Alaska: Evolution of a Continental Margin," *Geological Society of America Bulletin* 93 (1982):759–783.

42. D. Schlichter and G. Liebezeit, "The Natural Release of Amino Acids from the Symbiotic Coral *Heterozenia fuscescens* (Ehrb.) as a Function of Photosynthesis," *Journal of Experimental Marine Biology and Ecology* 150 (1991):83–90; D. Schlichter and H. W. Fricke, "Mechanisms of Amplification of Photosynthetically Active Radiation in the Symbiotic Deep-Water Coral *Leptoseris fragilis*," *Hydrobiologia* 216/217 (1991):389–394; D. Schlichter, H. W. Fricke, and W. Weber, "Light Harvesting by Wavelength Transformation in a Symbiotic Coral of the Red Sea Twilight Zone," *Marine Biology* 91 (1986):403–407; D. Schlichter, H. W. Fricke, and W. Weber, "Evidence for PAR-Enhancement by Reflection, Scattering and Fluorescence in the Symbiotic Deep Water Coral *Leptoseris fragilis* (PAR = Photosynthetically Active Radiation)," *Endocytobiosis and Cell Research* 5 (1988):83–94; H. W. Fricke, E. Vareschi, and D. Schlichter, "Photoecology of the Coral *Leptoseris fragilis* in the Red Sea Twilight Zone (an Experimental Study by Submersible)," *Oecologia* 73 (1987):371–381.

43. M. M. Littler, D. S. Littler, S. M. Blair, and J. N. Norris, "Deepest Known Plant Life Discovered on an Uncharted Seamount," *Science* 227 (1985):57–59.

44. There is clear evidence at Sekwi Brook for bacterial mat binding of the sediment. Considering that such bacterial mats can be of great lateral extent, there is no reason to believe that there was sufficient mud in suspension in this quiet water depositional environment to shut off photosynthesis on the seafloor. Even mud unbound by bacterial mats may not have contributed much to turbidity because the mudstones of this depositional environment were cohesive. The Canadian Ediacaran organisms were probably living in a quiet water (and hence lower turbidity) environment below fair-weather wave base, an environment with sufficient light to sustain photoautotrophy. Assuming that planktonic larval settlement was the dispersal mechanism of the Ediacaran organisms, ambient light levels may have been a main criterion for determination by larvae of substrate suitability.

Although it has been argued that trace fossils from the Blueflower Formation represent an assemblage indicative of the deepwater *Nereites* ichnofacies, these meandering burrows originated in shallow marine waters. The supposedly primitive version of phobotaxis in the trace fossil *Helminthopsis irregularis* is unconvincing because it represents a eurybathyic ichnospecies. These trace fossils are similar to many occurring in open shelf, photic zone marine depositional environments. The specimens of *Helminthoidichnites tenuis* from the Blueflower do not appear to show the phobotaxis typical for the ichnospecies; indeed, traces such as these could occur in any low-energy, oligotrophic aquatic environment below fair-weather wave base. In fact, *Helminthoidichnites tenuis* (as *Gordia marina*) from the shallow sublittoral deposits (with an Ediacaran body fossil biota) from the Wernecke Mountains is identical to the same ichnospecies from the Sekwi Brook biota. The Sekwi Brook specimens of *Planolites montanus, Torrowangea rosei, Palaeophycus tubularis*, and even *Helminthoidichnites tenuis* show frequent, and in the former two, abundant crossing of burrow tracks (G. M. Narbonne and H. J. Hofmann, "Ediacaran Biota of the Wernecke Mountains, Yukon, Canada," *Palaeontology* 30 [1987]:647–676), strongly suggesting an open shelf habitat with local relief rather than the *Nereites* ichnofacies. Geological data described by G. D. Delaney ("The Mid-Proterozoic Wernecke Supergroup, Wernecke Mountains, Yukon Territory," in F. H. A. Campbell, ed., *Proterozoic Basins of Canada*, pp. 1–24 [Canadian Geological Survey Paper 81–10, 1981]) also cast doubt on an assumption here of great depths—the middle Proterozoic rocks of the Wernecke Supergroup, which underlie the Windermere Supergroup, consist of presumed deeper-water sediment overlain by a 1.5-km-thick pile of shallow dolomite shelf sediment. Thus the Windermere is deposited at the end of a shallowing upward trend in the underlying sedimentary deposits.

The Canadian Proterozoic trace fossils of Narbonne and Aitken (1990) probably formed within an offshore part of the photic zone, and the Sekwi Brook biota should not be cited in support of the hypothesis that Ediacaran organisms lived below the photic zone.

45. See Vernadsky, 1997.

46. D. J. Nunkley, R. N. Jinks, B.-A. Battelle, E. D. Herzog, L. Kass, G. H. Renninger, and S. C. Chamberlain, "Retinal Anatomy of a New Species of Bresiliid Shrimp from a Hydrothermal Vent Field on the Mid-Atlantic Ridge," *Biological Bulletin (Woods Hole)* 190 (1996):98–110.

47. C. L. Van Dover, E. Z. Szuts, S. C. Chamberlain, and J. R. Cann, "A Novel Eye in Eyeless Shrimp from Hydrothermal Vents of the Mid-Atlantic Ridge," *Nature* 337 (1989):458–460; D. G. Pelli and S. C. Chamberlain, "The Visibility of 350°C Black-Body Radiation by the Shrimp *Rimicaris exoculata* and Man," *Nature* 337 (1989):460–461.

48. C. Zimmer, "The Light at the Bottom of the Sea," *Discover* 17 (1996):62–73.

49. C. Zimmer, 1996.

50. H. Walter, "The Namib Desert," in M. Evenari, I. Noy-Meir, and D. W. Goodall, eds., *Hot Deserts and Arid Shrublands*, pp. 245–282 (Amsterdam: Elsevier, 1985).

51. A. Seilacher, "Late Precambrian and Early Cambrian Metazoa: Preservational or Real Extinctions?" in H. D. Holland and A. F. Trendall, eds., *Patterns of Change in Earth Evolution*, pp. 159–168 (Berlin: Springer-Verlag, 1984).

52. McMenamin, 1986.

53. H. Felbreck, "Chemoautotrophic Potential of the Hydrothermal Vent Tube Worm, *Riftia pachyptila* Jones (Vestimentifera)," *Science* 213 (1981):336–338.

54. C. M. Young, E. Vázquez, A. Metaxas, and P. A. Tyler, "Embryology of Vestimentiferan Tube Worms from Deep-Sea Methane/Sulphide Seeps," *Nature* 381 (1996):514–516.

55. J. K. Rigby and J. L. Geodert, "Fossil Sponges from a Localized Cold-Seep Limestone in Oligocene Rocks of the Olympic Peninsula, Washington," *Journal of Paleontology* 70 (1996):900–908.

56. R. A. Fortey, "Review of M. A. S. McMenamin and D. L. S. McMenamin, 1990, *The Emergence of Animals: The Cambrian Breakthrough,* New York: Columbia University Press," *Historical Biology* 6 (1990):70–71.

57. M. McMenamin, "Osmotrophy in Fossil Protoctists and Early Animals," *Invertebrate Reproduction and Development* 22, nos. 1–3 (1993):301–304.

58. Vernadsky, 1997.

59. Y. Sugimura and Y. Suzuki, "A High Temperature Catalytic Oxidation Method for Determination of Non-Volatile Dissolved Organic Carbon in Sea Water by Direct Injection of Liquid Sample," *Marine Chemistry* 24 (1988):105–131.

60. G. A. Riley, "Particulate and Dissolved Organic Carbon in the Oceans," *United States Atomic Energy Commission Conference Report* 720510 (1973):204–220.

61. As reported on pp. 341–342 in S. W. Fox, ed., *The Origins of Prebiological Systems and of Their Molecular Matrices* (New York: Academic Press, 1965).

62. D. L. Meyer, "Ecology and Biogeography of Living Classes," in T. W. Broadhead and J. A. Waters, eds., *Echinoderms: Notes for a Short Course,* pp. 1–14 (Knoxville: University of Tennessee, 1980).

63. J. A. Karhu and H. D. Holland, "Carbon Isotopes and the Rise of Atmospheric Oxygen," *Geology* 24 (1996):867–870.

64. S. E. Frederick, P. J. Gruber, and N. E. Tolbert, "The Occurrence of Glycolate Dehydrogenase and Glycolate Oxidase in Green Plants: An Evolutionary Survey," *Plant Physiology* 52 (1973):318–323.

65. M. McMenamin, 1993.

66. D. Schlichter and G. Liebezeit, "The Natural Release of Amino Acids from the Symbiotic Coral *Heterozenia fuscescens* (Ehrb.) as a Function of Photosynthesis," *Journal of Experimental Marine Biology and Ecology* 150 (1991):83–90.

67. As opposed to mineral, and the distinction is important.

68. Mirovia is the world ocean of the Proterozoic world; see I. W. D. Dalziel, "Neoproterozoic-Paleozoic Geography and Tectonics: Review, Hypothesis, Environmental Speculation," *Geological Society of America Bulletin* 109 (1997):16–42.

69. Internal symbiotic microbes.

70. M. A. S. McMenamin and L. Bennett, "Light Transmission Through Sand: Implications for Photosynthetic Psammophiles and *Pteridinium,*" *Geological Society of America Abstracts with Program* 26 (1994):A-54.

71. E. Schulz, "Das Farbstreifen-Sandwatt und seine Fauna, eine ökologische-biozönotische Untersuchung an der Nordsee," *Meereskundliche Arbeiten der Universität Kiel* 3, no. 19 (1937):359–378.

72. C. Hoffmann, "Über die durchlässigkeit dünner sandschichten für licht," *Planta* 36 (1949):48–56.

73. Which, come to think of it, is a more convenient experimental setup than the one we had used, although if you want to test a variety of sand grain sizes, you will have to make a lot of glass sandwiches.

74. Manufactured by B. Lange in Berlin.

75. W. E. Krumbien, D. M. Paterson, and L. J. Stal, eds., *Biostabilization of Sediments* (Oldenburg, Germany: Bibliotheks-und Informationssystem der Universitaet Oldenburg, 1994).

76. See figures 4.2–4.5 in H. D. Pflug, "Role of Size Increase in Precambrian Organismic Evolution," *Neues Jahrbuch für Geologie und Paläontologie, Abhandlungen* 193, no. 2 (1994):245–286. Runnegar calls *Petalostroma kuibis* a probable pseudofossil; see p. 307 in B. Runnegar, "Vendobionta or Metazoa? Developments in Understanding the Ediacara 'Fauna,' " *Neues Jahrbuch für Geologie und Paläontologie, Abhandlungen* 195, nos. 1–3, (1995):303–318.

77. It is not always possible to properly assess this until the negatives are printed.

78. Sulfate is a common constituent of seawater.

79. Note in figure 5.5 how the X-ray image of the bedding plane curves downward as it approaches the chaperone wall. This is strong evidence that the chaperone wall was partly exposed and is in accord with the life position inferred for *Ventogyrus*. This new evidence was not in accord with the conclusions of T. P. Crimes and M. A. Fedonkin ("Biotic Changes in Platform Communities Across the Precambrian Phanerozoic Boundary," *Rivista Italiana di Paleontologia e Stratigrafia* 102 [1996]:317–332), who interpret a specimen of *Pteridinium* (in which the paired bathtubs are closed at the top to form paired tubes) as evidence that *Pteridinium* grew entirely within the sediment. In my opinion, this specimen may have been deformed by currents that folded the outer sides of the *Pteridinium*'s bathtubs over its top.

80. Sulfur-reducing bacteria are heterotrophs and thus require an outside energy source (in this case, carbon compounds). The pyrite they produce is analogous to the carbon dioxide we respire.

81. K. M. Schopf and T. K. Baumiller, "A Biomechanical Approach to Ediacaran Hypotheses: Weeding the Garden of Ediacara," *Geological Society of America Abstracts with Programs* 27 (1995):A-269.

82. See p. 126 and figure 2 in D. H. Erwin, J. W. Valentine, and D. Jablonski, "The Origin of Animal Body Plans," *American Scientist* 85 (1997):126–137.

83. B. M. Waggoner, "Phylogenetic Hypotheses of the Relationships of Arthropods to Precambrian and Cambrian Problematic Fossil Taxa," *Systematic Biology* 45 (1996):190–222.

84. D. Collins, "The 'evolution' of *Anomalocaris* and Its Classification in the Arthropod Class Dinocarida (nov.) and Order Radiodonta (nov.)," *Journal of Paleontology* 70 (1996):280–293.

85. The supposed eye ridges are invisible on all published photographs of *Bomakellia* and were not noticed by Fedonkin, who described the genus.

86. See especially plate 22, figure 1 in G. J. B. Germs, "The Stratigraphy and Paleontology of the Lower Nama Group, South West Africa," *University of Cape*

Town Department of Geology Chamber of Mines Precambrian Research Unit Bulletin 12 (1972):1–250. This was the specimen I was referring to as a "probable sprigginid fossil (possibly *Marywadea*)" in M. A. S. McMenamin, "A case for Two Late Proterozoic–Earliest Cambrian faunal province loci," *Geology* 10 (1982):290–292. Thus, although I was still holding to an animalian interpretation of Ediacarans, I was making a link between the "frondlike" fossils and the "wormlike" fossils.

87. G. J. B. Germs, "Possible sprigginid worm and a new trace fossil from the Nama Group, South West Africa," *Geology* 1 (1973):69–70. The putative sprigginid or marywadeid referred to by Martin Glaessner in the 1970's is probably this specimen (G. M. Narbonne, written communication, e.g., see reference to "segmented worm" on p. 120 in M. F. Glaessner, "Precambrian Paleobiology: Middle and Late Precambrian Life and Environment," in L. Motz, ed., *The Rediscovery of the Earth*, pp. 115–124 [New York: Van Nostrand Reinhold, 1975]).

88. G. M. Narbonne, B. Z. Saylor and J. P. Grotzinger, 1997, "The Youngest Ediacaran Fossils from Southern Africa," *Journal of Paleontology* 71 (1997): 953–967.

89. B. M. Waggoner and J. W. Hagadorn, "Ediacaran fossils from western North America: Stratigraphic and paleogeographic implications," *Geological Society of America Abstracts with Program* 29 (1997):A-30.

6 · *Cloudina*

Of the homogenous parts of animals, some are soft and fluid, others hard and solid; and of the former some are fluid permanently, others only so long as they are in the living body.

—Aristotle[1]

[The] evolution of the skeleton has clearly widened the possibilities for the development of life and thus added to the stream of biogenic migration.

—Vladimir I. Vernadsky[2]

It seems odd that there could be a sea in the middle of the California desert, but there it is. My first introduction to the Salton Sea came in the mid-1970s, when I visited its shores on a field trip. How did marine waters, with a marine fauna, end up in this unlikely spot?

By 1904 there were 100,000 acres under irrigation in California's Imperial Valley. Water was diverted from the Colorado River and directed to the Imperial Valley's agricultural zone. All was going well until silt plugged the head of the canal from the Colorado. Water delivery to farmers slowed to a trickle and legal action was threatened.

In fall of 1904 the California Development Company made a cut in the riverbank in an attempt to bypass the blockage. The diversion was poorly engineered, however, and the Colorado River surged completely out of control in the spring floods of 1905.[3] Freshwater rushed through the cut and into its old overflow channel, the Alamo River, and then plunged into the New River that flows north from the gulf. The flood downcut into the soft silts, creating a 28-ft waterfall and widening the New River to a width of a quarter mile. Not until February 1907 did a workforce of 2000 laborers and 3000 railroad cars carrying fill finally turn the Colorado back to the gulf.

Being an interior basin in an arid climate subject to irrigation, the water of the new sea became increasingly salty until it matched and later exceeded marine salinity. Now 35 × 15 miles, the Salton Sea was expected to dry up but did not because drainwater continued to flush into it from the irrigated fields of the Imperial Valley; by the late 1920s it was too salty for rainbow trout (*Salmo gairdnerii*); by 1950 state biologists succeeded with introductions of orangemouth corvina (*Cynoscion xanthulus*) and gulf croaker (*Bairdiella icistia*). With croakers as prey for the voracious corvina, a successful saltwater game fishery developed on what was once desert "wasteland."

Perhaps as a by-product of the introduction of marine fishes to this inland sea, barnacles were also introduced to the Salton Sea. After we reached the Salton Sea shore during my high school field trip, I was astonished to note that the strand line where we disembarked from our vehicle was composed of enormous numbers of fragmented barnacle shells. The shells were mostly broken down into individual platelets.

The remains of these sessile arthropods composed a coarse sand on the beach and were heaped into elongate sand bars parallel to the shore. I had never seen, and have never seen since, barnacle shells in this abundance on the California coast or anywhere else. Why were they so abundant here? Were the conditions for the growth of barnacles particularly favorable in the Salton Sea? Was there an absence of natural checks on population, such as predators, in this newly created habitat?

Years later, my experience on this desert shoreline proved valuable. A graduate student in search of a thesis topic, I had joined a joint United States-Mexican expedition to Sonora. The aim of the field work was to assess mineral resources (the primary Mexican interest) and to make stratigraphic correlations to rock units in the United States (the main U.S. Geological Survey interest). My main interest was to locate early shelly fossils or Ediacarans, about which I hoped to write my dissertation.

Jack Stewart, leader of the expedition and an expert Survey stratigrapher, was in his good-natured way giving me a hard time about my search for the fossils. He viewed it as somewhat quixotic, and told me so as we hiked up "Conophyton Canyon" in the Cerro Rajón. Unless he already knew more than I give him credit for, however, I had the last laugh that day, for after only a few hours of measuring stratigraphic section, we came upon a find of considerable importance. From my Field Notebook #4, an entry dated March 9, 1982 (figure 6.1):[4]

[We have come upon a bed of shelly fossils in the La Ciénega Formation] associated with a cross-bedded conglomeratic horizon. This may be a paleontologic find of some importance. Jack found it first while we were measuring section. These appear to be shelly fossils, tubular or conical with a circular cross-section, and

56

╬

The fossils are silicified, and crop out in a 20m
laterally continuous ~~coarse sand,~~ granule to cobble conglomeratic unit
that seems to pinch out (or be eroded off?) to the west. Preservation
varies from poor to good, fossils weather out in
relief.

Siltstones &
Shales
MM - 82 - 43

Shelly? beds
are associated
with a crossbedded
conglomeratic horizon   ⅡC 1,2

This may be a paleontologic find of some importance. Jack found
it first while we were measuring section. These appear to be
shelly fossils, tubular or conical with a circular cross-section, and
with annualtions on the outer surface. The wall appears to
be double in some examples:

to 5mm
1-3 mm

Some of the tubes appear smooth-walled & tapered

FIGURE 6.1: An entry from my Field Notebook #4 dated March 9, 1982. This entry
records Jack Stewart's and my discovery of shelly fossils in the La Ciénega Formation of
the Cerro Rajón, Sonora, Mexico.

with annulations on the outer surface. The wall appears to be double in some examples.

Some of the tubes appear smooth-walled and tapered.

I could scarcely believe my good fortune at this moment, having what appeared to be the fossils I was looking for, and in great abundance. Tracing the fossiliferous bed laterally led to places where the shells were piled into beds a meter thick.

The fossils looked somewhat strange, however, not like ordinary fossils, and for some moments I debated whether they were fossils. Some of the specimens appeared to be melded into others as if they were made of soft plastic.

The beds Jack and I had found had been noted before by Allison "Pete" Palmer, a famous paleontologist in the Cambrian professional circle who, perhaps because of their unusual melded appearance, had interpreted them as dolomitized (that is, crystal-hardened) burrows. This was not an unreasonable interpretation considering the type of reconnaissance work he was engaging in. I now believe that the plastic appearance of some of the shells was caused by postburial alteration of the sediments in which the shells were preserved, and associated smashing of the fossils.

The shells were mostly fragmentary and constituted a rock type known as a coquina. *Coquina,* a geological term taken from the Spanish diminutive term for *shellfish* (itself derived from the Latin word *concha*), refers to a sedimentary rock, often limestone, composed almost entirely of shells and shell fragments.

Some reflection on the find convinced me that they were indeed shelly fossils (figure 6.2), and something seemed familiar about their mode of occurrence. Of course! These were nearshore specimens, broken and heaped into coarse sand bars just like the barnacle shells of Salton Sea. If the Salton Sea barnacles were to fossilize, they would produce a new coquina similar to this ancient one.

I authored or coauthored with Jack Stewart and others several scientific papers describing the shelly fossils, and the find became the centerpiece of my dissertation. Other scientists took notice of the find, which led me to a number of highly informative scientific contacts. Using the fossils (members of the genus *Sinotubulites*), I made highly accurate correlations both to fossiliferous rocks in the United States (thus helping to fulfill the original mission of the U.S. Geological Survey) and to rocks in China. Last but not least, I received a Ph.D. for my efforts.

The Mexican *Sinotubulites* fossils, which I described as the species *Sinotubulites cienegensis*,[5] belong to a group of fossils known as cloudinids (family Cloudinidae). In a situation with an odd parallel to Trevor Ford's original description of the Ediacaran frond fossil *Charnia,* cloudinids were first described as a type of alga by Beurlen and Sommer in 1957.[6] In the middle part of this century, "alga" was apparently the default taxonomic assignment for indeterminate ancient fossils.

(a)

(b)

FIGURE 6.2: The cloudinid shelly fossil *Sinotubulites cienegensis* from the La Ciénega Formation of the Cerro Rajón, Sonora, Mexico. (a) View of the sandy dolostone in which the fossils occur. (b) Enlarged view of several of the *Sinotubulites* fossils. Width of view approximately 4 mm.

The first cloudinid locality was discovered in Brazil by Dr. Octávio Barbosa from a limestone quarry at Ladário near Corumbá in the state of Mato Grosso. This locality was later recollected by Dr. Luciano Jacques de Morais. Jacques de Morais correctly interpreted the specimens as fossils of animal origin and then gave the specimens to Beurlen and Sommer for formal description. Beurlen and Sommer linked the specimens to *Aulophycus,* a genus of Cambrian calcareous algae, and named the Brazilian fossil *Aulophycus lucianoi* in honor of Luciano Jacques de Morais. In 1984, Zaine and Fairchild determined that the fossils did not in fact belong to the algal genus *Aulophycus,* thus returning to Luciano Jacques de Morais's original interpretation.[7] Zaine and Fairchild noted close similarities to the Namibian genus *Cloudina,* and transferred the Brazilian species to this genus as *Cloudina lucianoi* (Beurlen and Sommer 1957).

North American cloudinids were first described as *Wyattia* by Taylor, who was unaware that he had fossils similar to the ones from Brazil.[8] In Africa, Germs named the genus *Cloudina,* after Preston Cloud, as part of his work on platy limestones of the African Nama Group in the 1970s.[9] Hahn and Pflug found more cloudinids in South America, and established the new family Cloudinidae in 1985.[10] An exceptional possible occurrence of cloudinid fossils is known from carbonate rocks of the Stirling Formation of eastern California.[11] If confirmed as cloudinids, these would be the oldest cloudinids known.

Cloudinids are tubular fossils up to several centimeters in length (figure 6.3). Shell structure in *Cloudina* consists of tube wall layers arranged in cone-in-cone fashion, like stacks of Styrofoam cups with their bottoms stretched downward. Each individual calcareous wall layer in a cloudinid skeleton is thin and delicate, but it tends often to be made more robust by multiple, closely nested, adjacent walls and by secondary crystallization of calcite and dolomite crystals around and between the thin shell walls. This permineralization apparently occurred after the death of the organism. Despite the fact that the shell has multiple wall layers, each separated by a narrow space, the cloudinid shell was not built to stand up to much punishment, either mechanical stresses or biological attack. Indeed, as I first pointed out, cloudinids (particularly those occurring in coquinas) are often found with their outer walls stripped off by mechanical breakage, leaving behind a cylindrical core.[12]

It now appears as if all the cloudinids are closely related. Three genera can be defined:

Wyattia: Classic *Wyattia* and Langille's[13] form from Death Valley, California; generally smaller than the other two genera
Sinotubulites: Cloudinids from China, Mexico, Nevada, and Oman
Cloudina: cloudinids from Brazil, Namibia, and Spain

All three families belong to the genus Cloudinidae. *Wyattia* seems to be characteristic for North America and may be the oldest cloudinid or

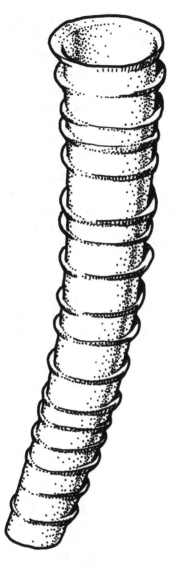

FIGURE 6.3: A cloudinid, the earliest known shelly fossil that may be a calcareous shell of an animal. Cloudinids are known from Namibia, Brazil, Spain, Mexico, and elsewhere. Length of specimen 2 cm.

From M. A. S. and D. L. S. McMenamin, *The Emergence of Animals: The Cambrian Breakthrough* (New York: Columbia University Press, 1990). Artwork by Dianna McMenamin.

ancestral to the other genera. This would accord with the recent recognition of the oldest Ediacarans in western North America.[14] *Cloudina* is more widespread, occurring in Brazil, Namibia, and Spain. It has been argued that these different genera are merely preservational variants of the same genus,[15] but the scientific opinion seems to be converging on the view that there was more than one genus of cloudinids.

The first question that now must be asked of cloudinids is, "Do they represent the shells of animals?" Recall that in the original description they are described as calcareous marine algae. Could this view still be correct? Or is it merely a legacy of old-fashioned, anthropocentric thought, as engaged in by A. C. Hardy in 1953, when he saw the "primitive" plants as ancestral to the more "advanced" animals?[16] Indeed, true fossil algae with calcareous skeletons are known from the same strata that bear the cloudinids. Glaessner, true to his preferences for the interpretation of ancient fossils as animals, saw the cloudinids as fossils of tube-dwelling annelid worms.[17]

The second question to be asked of cloudinids is, "Why do they have shells?" Cloudinids are without question the first animal-style shelly organisms to occur abundantly in the fossil record.[18]

In 1992 I was invited to participate in a symposium held during the 129th annual meeting of the National Academy of Sciences in Washington, D.C., honoring the life and career of Preston Cloud. The symposium was titled "Where, When, and How Did Life Originate? A Discussion" and was held on April 26 at the prestigious National Academy of Sciences center. As part of the discussion, I sketched the main outlines of the Garden of Ediacara theory.

Shortly after speaking I was challenged from the audience by Stephan Bengtson, a Swedish paleontologist interested in Cambrian fossils. Bengtson claimed to have found as yet unpublished boreholes in cloudinid specimens from China and also claimed that this overturned the Garden of Ediacara theory because one of its tenets is that large predators were rare or absent in the Precambrian.[19] I replied that assuming Bengtson's results were valid, the boring organisms were very small and not necessarily a threat to all multicellular organisms of the time in the same way as would be, say, an *Anomalocaris* specimen with eyes, ears, a nose, and so on. The audience liked this and some in the audience later told me that I had had the better of the exchange.

More recent work indicates that Bengtson's boreholes may simply be damage to shell caused after the inhabitant was deceased. But these holes could still represent an important discovery, and Bengtson's claim is worthy of further evaluation.

Bengtson's borer, if it existed, may have had eyes or it may have been blind or in possession of only very rudimentary eyepatches. A piece of negative evidence favoring the Garden of Ediacara hypothesis is the absence of convincing eyes in the Proterozoic fossils. How much geological time would it require to evolve a sharply focusing, image-making eye from, say, a flat skin patch of light-sensitive skin?

Early evolutionary debates centered on how an organ such as the octopus eye could evolve from less complex ancestral organs. Dan Nilsson and Susanne Pelger recently estimated the required amount of time, and it turned out to be shorter than one might expect.[20] In their computer model, they began with a flat patch of light-sensitive skin covered by a transparent protective layer. The computer model simulated evolution by allowing the eye shape to mutate, the only constraint being that change from one generation to the next could be only 1 percent different in either direction (for example, larger or smaller) than what had come before. The results were dramatic. From its flat beginning, the simulated eye "evolved" steadily to a shallow cup-shaped retina and from there to an increasingly spherical eyeball. The most impressive part of the study is that at each step along the way, the evolving eye experienced measurable improvement in visual acuity.

Nilsson and Pelger estimated that the time required to evolve from a flat light-sensitive patch to an eyeball was less than 400,000 generations. This, in terms of vision, is the same as going from blind flatworm to *Anomalocaris* in 500,000 years.

Animals such as small fish typically have generation times of 1 year or less, so the evolution from the eyepatch state to the eyeball state could be accomplished in less than a half million years. As put by one prominent evolutionist, this is an interval of time usually too short for geologists to measure, and is thus a mere geological blink. The implications of such an evolutionary rate of change for our understanding of the apparent suddenness of the Cambrian explosion are great, as are the evolutionary advantages that would have accrued to organisms with shells once the vision-directed Cambrian predators made their appearance. I am not convinced that cloudinids were already experiencing the onslaught of the scary creatures with eyes, but it remains an intriguing possibility.

Notes

1. *Parts of Animals*, Book 2:2, p. 51 in Aristotle, *On Man in the Universe* (Roslyn, N.Y.: Walter J. Black, 1943).

2. V. I. Vernadsky, *The Biosphere.* Translated by D. Langmuir, revised by M. McMenamin. (New York: Nevraumont/Copernicus, 1997).

3. R. H. Boyle, "Life or Death for the Salton Sea?" *Smithsonian* 27, no. 3 (1996):87–96.

4. Page 56.

5. M. A. S. McMenamin, "Basal Cambrian Small Shelly Fossils from the La Ciénega Formation, Northwestern Sonora, Mexico," *Journal of Paleontology* 59 (1985):1414–1425.

6. K. Beurlen and F. W. Sommer, "Observaçoes estratigráficas e paleontólogicas sôbre o calcário Corumbá," *Boletim, Departamento Nacional da Producao Mineral* 168 (1957):1–35.

7. M. F. Zaine and T. R. Fairchild, "Comparison of *Aulophycus lucianoi* Buerlen and Sommer from Ladário (MS) and the Genus *Cloudina* Germs, Ediacaran of Namibia," *Anais da Academia Brasilera de Ciências* 57 (1985):130.

8. M. E. Taylor, "Precambrian Mollusc-Like Fossils from Inyo County, California," *Science* 153 (1966):198–201.

9. G. J. B. Germs, "New Shelly Fossils from the Nama Group, South West Africa," *American Journal of Science* 272 (1972):752–761.

10. G. Hahn and H. D. Pflug, "Die Cloudinidae n. fam., Kalk-Röhren aus dem Vendium und Unter-Kambrium," *Senckenbergiana lethaea* 65 (1985):413–431.

11. G. B. Langille, "Problematic Calcareous Fossils from the Stirling Quartzite, Funeral Mountains, Inyo County, California," *Geological Society of America Abstracts with Program* 6 (1974):204–205.

12. McMenamin, 1985.

13. Langille, 1974.

14. M. A. S. McMenamin, "Ediacaran Biota from Sonora, Mexico," *Proceedings of the National Academy of Sciences USA* 93 (1996):4990–4993.

15. S. Conway Morris, B. W. Mattes, and Chen Menge, "The Early Skeletal Organism *Cloudina*; New Occurrences from Oman and Possibly China," *American Journal of Science* 290-A (1990):245–269; S. W. F. Grant, "Shell Structure and Distribution of *Cloudina*: A Potential Index Fossil for the Terminal Proterozoic," *American Journal of Science* 290 (1990):261–294.

16. A. C. Hardy, "On the Origin of the Metazoa," *Quarterly Journal of the Microscopical Science* 94 (1953):441–443.

17. M. F. Glaessner, "Early Phanerozoic Annelid Worms and Their Geological and Biological Significance," *Journal of the Geological Society of London* 132 (1976):259–275.

18. Fossil skeletons of sponges or spongelike organisms may be of similar antiquity. See figure 2A in J. P. Grotzinger, S. A. Bowring, B. Z. Saylor, and A. J. Kaufman, "Biostratigraphic and Geochronologic Constraints on Early Animal Evolution," *Science* 270 (1995):598–604.

19. Later published as S. Bengtson and Yue Zhao, "Predatorial Borings in Late Precambrian Mineralized Exoskeletons," *Science* 257 (1992):367–369.

20. D.-E. Nilsson, "Vision, Optics and Evolution," *Bioscience* 39 (1989):298–307; R. Dawkins, "The Eye in a Twinkling," *Nature* 368 (1994):690.

7 · *Ophrydium*

I haven't convinced everybody, and I know it will be a while. There
will be tweakings, certain modifications as we learn more.

—Lynn Margulis[1]

In 1985, I was contemplating the scientific work of Lynn Margulis, who
at the time was at the other end of the state at Boston University. I was
delighted to learn that she was in the process of being hired at the
University of Massachusetts at Amherst, just 12 miles to the north of
Mount Holyoke. After her arrival, I attended her lecture at Hampshire
College and introduced myself to her afterward. We had actually met
years before in the Clean (now Cloud) Laboratory at the University of
California at Santa Barbara, but now, being at adjacent institutions, we
had an opportunity to interact.

Margulis is a fascinating person, often delightedly full of plans and
schemes for the next breakthrough research program or major field
expedition. At all times she manifests a surging undercurrent of scientific brilliance. Sometimes the current bursts forth, and she regales students and associates with streams of her scientific insight, for she ascribes
to a heuristic mission that she takes as seriously as her research and theoretical pursuits. And she takes these seriously enough; as put by one science writer, "Margulis's batting average in the big-theory game is better
than Ty Cobb's with men on base."[2]

Margulis's involvement with Ediacarans is manifest in her study,
undertaken with student Brian Duval, of the protist *Ophrydium*.
Ophrydium is a sessile ciliate of the order Peritrichida. This microbe
forms huge (up to 15 cm long), green, gelatinous colonies of individual
ciliate microbes. These colonies live in bog wetlands, such as Hawley
Bog and Leverett Bog in Massachusetts. They also occur in brackish
coast ponds of the Arctic. The colonies, when brought to the surface
with a dip net, resemble greenish balls of jelly.

Each *Ophrydium* individual consists of an elongate, cylindrical cell
tapered at one end (color plate 14). Each cell is called a zooid. First
described in 1786, every zooid in the colony resembles a tiny eyebrow
or perhaps eyebrow hair when viewed in a light microscope, hence the

genus name taken from the Greek word *ophrys* (eyebrow). The tapered end of the cell is tethered by a thin fiber, which anchors the individual to the rest of the colony. This fiber is called a myoneme, or "muscle thread." The microbes radiate outward in a centripetal fashion to form the mass of the colony. Each *Ophrydium* cell is embedded in a polysaccharide gel; this gel forms the mass of the colony (color plate 15).

The green color of the jelly ball is imparted by symbiotic eukaryotes living within the *Ophrydium* cells. Each individual *Ophrydium* is packed with endosymbiotic microbes. These microbes are unidentified as of yet, but they resemble the unicellular green alga *Chlorella.*

The entire jelly glob, then, is a microbial photosynthesis factory. A host of other organisms, besides *Ophrydium,* also make the jelly glob home. Duval and Margulis have cataloged three species of spirochete bacteria, a *Saprospira*-like bacterium, numerous rod-shaped bacteria, more than four types of cyanobacteria, heliozoans, two species of desmids, more than four types of green algae, six genera of diatoms, *Paramecium bursaria* (itself with algal endosymbionts), a ciliate protist (with endosymbiotic methanogenic bacteria), at least three other types of ciliates, a zoomastiginid protist, several other types of protists, cladoceran animals, copepods, rotifers, nematodes, and flatworms living within the gel in addition to *Ophrydium* and its algal endosymbiont. Clearly this is a complex association of organisms living in the nutritious mucus of the gel.

In my office at Mount Holyoke College, Brian Duval used the light transmission apparatus (color plate 16) designed by Lori Bennett to measure the light transmission through live *Ophrydium.* He found that 30–46 percent of the incident light from our fiber-optic light source was attenuated by a 1- to 2-cm thickness of *Ophrydium* gel.

Margulis and Duval maintain that *Ophrydium* might be a modern analog for the marine Ediacaran biota. And certainly it could make a very nice partial ecological analog to the ancient Garden of Ediacara habitat. Unfortunately, *Ophrydium* has no fossil record and is unlikely to have one because jelly balls have an extremely slim likelihood of preserving as recognizable fossils. Nevertheless, it *is* likely that colonies of organisms like *Ophrydium* were alive during Ediacaran times; at the very least they provide a modern analogue for what might have been going on at that time. In the words of Duval and Margulis,

In physiological terms, the gel is both a "skeleton" and a "tissue" matrix of an individual *Ophrydium* mass. *Ophrydium* gel masses show properties of cohesion, coherence and integrity that are reminiscent of the loose individuality of certain large protoctists such as coralline algae and kelp. In eukaryotes individuality is always a complex product of the interaction of formerly independent individuals. We suggest that the loosely defined individuality of the *Ophrydium* colony is similar to that of the ancient Ediacaran biota. The "pneu" structural elements in the enigmatic late Proterozoic "garden of Ediacara" biota (e.g., *Pteridinium, Phyllozoon*) are possible ancient analogs to *Ophrydium* (McMenamin and McMenamin, 1990). During the Vendian period supplemental nutrition by phototrophy via microbial symbionts was likely in translucent large protoctists and/or early animals residing in the photic zone (McMenamin, 1993). The potential for nutrient and genetic exchange within an *Ophrydium*-like gel may have been optimal.[3]

The preceding excerpt, in which individuality is seen as a complex product of the interaction of formerly independent symbionts and endosymbionts, is a clear statement of the symbiogenesis worldview. These types of organizations of cells have been called an expression of "metacellularity."[4]

Unfortunately, the assertion that *Ophrydium*-like colonies lived during the Proterozoic is difficult to test using the fossil record. Paleontologists dream of a fortunate discovery of silicified late Precambrian seafloor, a chert deposit that preserves the microstructure of microbial communities and colonies directly associated with the Ediacaran biota. But such a find may be a long time coming.

Is there any evidence for coloniality among the Ediacaran forms? Glaessner[5] interpreted the spindle-shaped form (informally known as Vendofusa;[6] figure 7.1) as a possible hydrozoan colony, but this could not be if, as I suspect, the spindle-shaped form represents a single individual. However, there is evidence that Ediacaran forms were gregarious. Both *Pteridinium* and *Ernietta* in Namibia, and the spindle-shaped form in Newfoundland, form locally abundant clusters of individuals.[7]

The discoidal fossil *Beltanelliformis brunsae* (figure 7.2) occasionally occurs in great abundance on individual bedding planes.[8] Specimens are often found clustered together on a single bedding plane, and the size distributions on individual bedding planes are strongly unimodal, suggesting that each bedding plane sample represents a population of individuals at the same stage of growth. This might be taken as evidence that *Beltanelliformis* had planktonic larvae, in which an entire spatfall of related individuals would simultaneously colonize the seafloor surface.

FIGURE 7.1: The spindle-shaped Ediacaran, informally known as Vendofusa, from the Ediacaran locality of the Mistaken Point Formation, Conception Group, eastern Newfoundland. Scale bar in centimeters.

With regard to *Ophrydium,* no one knows how it reproduces, what its propagules look like, or how it is transported from one pond habitat to the next. We need to know this.

In spring 1997 Margulis informed me of a paper by M. Bauer-Nebelsick, C. F. Bardele, and J. A. Ott that had recently come to her attention.[9] Lynn mentioned that the protoctist described in this paper had an uncanny resemblance to certain members of the Ediacaran biota.

I was indeed impressed with these resemblances. My first opportunity to read the entire paper was on an airline flight with Lynn. We were en route to Mexico to visit the *humidales* (mangrove wetlands) of the west coast in search of colonial ciliates (which we did eventually find) and then to present lectures in Spanish for the University of Nuevo Leon in Monterrey.

As the title of the Bauer-Nebelsick et al. paper indicates, *Zoothamnium niveum* is a colonial ciliate coated with chemoautotrophic bacteria. In a striking resemblance to *Charniodiscus* from the Ediacaran biota, *Zoothamnium niveum* has a circular adhesive holdfast, an elongate stalk, and alternating branchlets connected along a zigzag medial suture (called the continuous spasmoneme in *Zoothamnium niveum*).

FIGURE 7.2: *Beltanelliformis brunsae,* a globular Ediacaran from northwestern Canada. The fact that all four of these specimens are virtually the same diameter (29 mm) suggests that they form a related cohort, perhaps part of the same spatfall. Sketch from epoxy cast.

However, the most interesting aspect of *Zoothamnium niveum* is that its branchlets (each of which is composed in turn of bell- or club-shaped zooids) are coated with a dense population of chemosymbiotic bacteria. The bacterial coating is sufficiently thick to give *Zoothamnium niveum* a whitish color. The spherical macrozooids of the colony (which eventually break away from the colony axis and function as propagules) reach up to 1.5 mm in diameter; the colony itself is visible to the naked eye, with some examples reaching lengths of several centimeters or more.

This occurrence of a large, modern colonial protoctist with Garden of Ediacara-style symbiotic bacteria has major implications for our understanding of the Ediacarans. Two possibilities immediately present themselves. First, *Zoothamnium niveum* supports the proposition that most or all of the Ediacaran body fossils were not animals and were perhaps colonial protoctists.

Few would argue that *Zoothamnium niveum* is a direct descendant of the Ediacaran body fossils. The second possibility, therefore, is that the resemblances between *Zoothamnium niveum* and the Ediacarans (be they protoctists, animals, or none of the above) are the result of a particularly impressive example of convergent evolution of a distinctive body form, well suited to the Garden of Ediacara trophic strategies of chemosymbiosis, photosymbiosis, and direct absorption of nutrients from water.

Notes

1. E. Royte, "Attack of the Microbiologists," *The New York Times Magazine* January 14, 1996:21–23.
2. See p. 44 in D. Quammen, "Sea and Hypersea: Staying Anchored in an Ocean of Big Thinking," *Outside* 10, no. 4 (1995):43–50.
3. See p. 199 in B. Duval and L. Margulis, "The Microbial Community of *Ophrydium versatile* Colonies: Endosymbionts, Residents and Tenants," *Symbiosis* 18 (1995):181–210.
4. See p. 87 in H. R. Maturana and F. J. Varela, *The Tree of Knowledge: The Biological Roots of Human Understanding* (Boston: Shambhala, 1987).
5. See p. A85, figure 3, in M. F. Glaessner, "Biogeography and Biostratigraphy—Precambrian," pp. A79–A118 in R. A. Robinson and C. Teichert, eds., *Treatise on Invertebrate Paleontology, Part A: Introduction, Fossilization (Taphonomy), Biogeography and Biostratigraphy* (Boulder, Colo., and Lawrence, Kans.: The Geological Society of America and the University of Kansas, 1979).
6. This fossil has not been formally named and will be informally referred to in this book as Vendofusa.
7. R. Bendick, "Spatial Statistics of the Earliest Metazoa: Simultaneous Rise of Complex Ecology and Morphology During the Vendian," *Geological Society of America Abstracts with Program* 26 (1994):A-54.
8. M. A. S. McMenamin, "The Dawn of Animal Life," *Episodes* 11 (1988):229–230; G. M. Narbonne and H. J. Hofmann, "Ediacaran Biota of the Wernecke Mountains, Yukon, Canada," *Palaeontology* 30 (1987):647–676. *Beltanelliformis* has been recently interpreted as a sand-filled body fossil with possible psammocoral affinities: A. Seilacher and R. Goldring, "Class Psammocorallia (Coelenterata, Vendian Ordovician): Recognition, Systematics, and Distribution," *Geologiska Föreningens i Stockholm Förhandlingar* 118 (1996):207–216.
9. M. Bauer-Nebelsick, C. F. Bardele, and J. A. Ott, "Redescription of *Zoothamnium niveum* (Hemprich & Ehrenberg, 1831) Ehrenberg, 1838 (Oligohymenophora, Peritrichida), a Ciliate with Ectosymbiotic, Chemoautotrophic Bacteria," *European Journal of Protistology* 32 (1996):18–30; M. Bauer-Nebelsick, C. F. Bardele, and J. A. Ott, "Electron Microscopic Studies on *Zoothamnium niveum* (Hemprich & Ehrenberg, 1831) Ehrenberg, 1838 (Oligohymenophora, Peritrichida), a Ciliate with Ectosymbiotic, Chemoautotrophic Bacteria," *European Journal of Protistology* 32 (1996):202–215.

8 · Reunite Rodinia!

> The theory of Wegener [continental drift] is to me a beautiful dream, the dream of a great poet. One tries to embrace it and finds that he has in his arms but a little vapor or smoke; it is at the same time both alluring and intangible. —Pierre Termier[1]

> We have known since the days of Kant that scientific arguments must never be founded on analogies, but the authors are dead serious about these poetic digressions. —Peter Westbroek[2]

> The continental land drift continued; increasingly the ocean penetrated the land as long fingerlike seas providing those shallow waters and sheltered bays which are so suitable as a habitat for marine life . . . [with] the further separation of the land masses and, in consequence, a further extension of the continental seas . . . these inland seas of olden times were truly the cradle of evolution. —*The Urantia Book*[3]

The last quotation in this chapter's epigraph describes the Proterozoic breakup of the supercontinent Rodinia. This amazing passage, written in the 1930s, anticipates scientific results that did not actually appear in the scientific literature until many decades later. This unusual source is *The Urantia Book.*[4] The name *Urantia* refers to planet Earth.

Like the *Book of Mormon* and L. Ron Hubbard's *Dianetics, The Urantia Book* is a modern attempt to found a new religion. But the teachings of *The Urantia Book,* as promoted by the Urantia Foundation and the Urantia Brotherhood,[5] are more mainstream than either Mormonism or dianetics. Promotional literature of the Urantia organization inserted into new copies of the book state the following:

> We hope your experience with the URANTIA teachings will enhance and deepen your relationship with God and your fellow man, and provide renewed hope, comfort, and reassurance in your daily life.

What more could one ask for in a religion? Well, for starters, one could hope for accurate geology and profound scientific truths in its sacred literature, something both the devout and the skeptics alike find lacking in much of the Bible.

The comments concerning Rodinia's breakup and its influence on animal evolution are found in part III, "The History of Urantia" in *The Urantia Book*. According to the first page of this chapter, "these papers were sponsored by a Corps of Local Universe Personalities acting by authority of Gabriel of Salvington." The critical section 8 of Paper 57, titled "Crustal Stabilization, The Age of Earthquakes, The World Ocean and the First Continent," is "presented by a Life Carrier, a member of the original Urantia Corps [who visited our planet hundreds of millions of years ago] and now a resident observer." The following Paper 58, "Life Establishment on Urantia," is attributed to "a member of the Urantia Life Carrier Corps now resident on the planet."

Clearly we are not dealing here with an orthodox scientific treatise. Nevertheless, the anonymous members of the Urantia Corps hit on some remarkable scientific revelations in the mid-1930s. They embraced continental drift at a time when it was decidedly out of vogue in the scientific community. They recognized the presence of a global supercontinent (Rodinia) and superocean (Mirovia), in existence on earth *before* Pangea. From *The Urantia Book*:

> 1,000,000,000 years ago . . . [t]he first continental land mass emerged from the world ocean. . . . 950,000,000 [years ago] . . . presents the picture of one great continent of land and one large body of water, the Pacific Ocean.[6]
>
> 800,000,000 years ago . . . Europe and Africa began to rise out of the Pacific depths along with those masses now called Australia, North and South America, and the continent of Antarctica, while the bed of the Pacific Ocean engaged in a further compensatory sinking adjustment. By the end of this period almost one third of the earth's surface consisted of land, all in one continental body.[7]

Of course I am being selective here in my choice of quotations, and there are reams of scientifically untenable material in *The Urantia Book*. However, the concept of a billion-year-old supercontinent (the currently accepted age for the formation of Rodinia) that subsequently split apart, forming gradually widening ocean basins in which early marine life flourished, is unquestionably present in this book.

Orthodox scientific arguments for such a proposal did not appear until the late 1960s, and a pre-Pangea supercontinent was never described until Valentine and Moores made the attempt in 1970. The Urantia Corps not only had the age of the formation of Rodinia approximately correct at 1 billion years, but they also were first to link breakup of Rodinia to the emergence of animals (even if the mode of appearance was implantation by extraterrestrials). Furthermore, they even got the timing of *that* approximately correct at 650 to 600 million years ago ("These inland seas of olden times were truly the cradle of evolution").[8]

This book was unknown to me until it was brought to my attention by J. J. Johnson in October 1995. I obtained a copy of the book from the Smith College library and noted the 1955 (eighth edition 1984) publication date. What could possibly explain such precocious insight from such an unexpected corner? Perhaps it has to do with a lively, unconstrained, but nevertheless informed imagination. John K. Wright has noted how outrageous hypotheses "arouse interest, invite attack, and thus serve useful fermentative purposes in the advancement of geology."[9] But what about outrageous religions?

I wrote back to Johnson on January 15, 1996, asking him whether he could confirm that the passages he had referred me to were indeed written in 1955. In a letter dated January 24, he replied that the section of interest was "put into the English language in 1934," making it even more ahead of its time than I had thought.

Johnson congratulated me on my fossil discovery south of Tucson (see chapter 9) and for my "appreciation for the Truth." He then invited me to contact the Fellowship for Readers of *The Urantia Book*. He gave me a contact address, telephone number and fax for the Fellowship and advised me to contact John Hales and to consider attending an event called the Flag Conference. I consider Johnson's (unsuccessful) attempt to convert me to his religion to be a very friendly overture, and although I cannot become a Urantia proselyte, I wish the members of this faith all the best.

Assuming for the moment that space voyagers are not responsible for life's origin and history on this planet, one wonders how the *Urantia Book* authors arrived at the concept of a Proterozoic supercontinent, and the link between breakup of this supercontinent and the emergence of complex life in the ensuing rift oceans, 30 years before most geologists accepted continental drift and nearly four decades before scientists had any inkling that Rodinia existed. The anonymous authors responsible for the critical part of section 3 evidently possessed a high level of geological training, and while writing in the 1930s must have known of Wegener's ideas on continental drift. Perhaps he or she was, or had contact with, an expatriate from Nazi Germany. Whatever the identity of the author, this person proceeded to speculate about the relationship between evolutionary change and the breakup of a Proterozoic supercontinent in an exceptionally fruitful way. Perhaps this was because the thought and the writing of this person were not fettered by the normal constraints of the (too often highly politicized) scientific review process.

Cases such as this one (which is by no means unique) are an exercise in humility for me as a scientist. How can it be that discovery of Rodinia, *plus* a fairly sophisticated rendering of the evolutionary implications of the rifting of Rodinia, falls to an anonymous author engaging in a work of religious revelation decades before scientists find out *anything* about the subject? Perhaps this is an important aspect of religion—a creative denial of certain aspects of reality in order to access a deeper truth.

I am not advocating an abandonment of a disciplined scientific peer review process, but I can't help but wonder whether science would benefit by having scientists themselves or friends of science systematically scan the various nonscientific literatures for writings such as those appearing in *The Urantia Book*. Scientists would ordinarily ignore and dismiss such writings, but a discerning eye might pick up some gems.

The concept of Rodinia therefore has a shockingly unexpected intellectual pedigree. When does the concept finally enter the conventional scientific channels? In articles published in the early 1970s, James W. Valentine and Eldridge M. Moores traced the geological history of the continents and spoke of a Precambrian supercontinent.[10] This continent was subsequently called proto-Pangea, pre-Pangea, Pangea I, the Late Proterozoic Supercontinent, ur-Pangea, or simply the Precambrian supercontinent. While writing *The Emergence of Animals*, Dianna McMenamin and I grew weary of these cumbersome names and proposed the name *Rodinia* for the ancient supercontinent. The corresponding superocean also needed a name, and we decided to call it Mirovia. Here is the key passage from *Emergence of Animals*[11]:

> Mirovia is derived from the Russian word *mirovoi* meaning "world or global," and, indeed, this ocean was global in nature. Rodinia comes from the infinitive *rodit*, which means "to beget" or "to grow." Rodinia begot all subsequent continents, and the edges (continental shelves) of Rodinia were the cradle of the earliest animals.

Curiously, *The Urantia Book* also refers to Mirovia, the "world ocean."[12] Here are my notes regarding the name from p. 17 of my 1987 composition notebook:

5/12/87 This book would be a good opportunity to "name" "paleo-Pangaea" and "proto-Panthallasa"

How about:

Ur-something

Rodinia from Russian *rod*: genus *rodit*: beget, come up, grow

Eomaria

Paleomaris

Mirovian Ocean from Russian *mirovoj*: World, Global, see pp. 19–20

[*the entry on composition notebook pp. 19–20 follows:*]

5/21[/87] Fedonkin, "Organicheskii Mir Venda" 1983 1210 pp. 4–5.

The glaciation at the beginning of the Vendian period, known under the name of the Laplandian or Varangian Glaciation, may have had catastrophic results for many groups of the organic world which inhabited the world ocean. (translation M. McMenamin, 5/21/1987)

As correctly pointed out by John J. W. Rogers, the word *Rodinia* is also derived from the Russian word *rodina*, meaning "motherland."[13] The term links the northern and southern hemispheres as well because of its phonic similarity to the Precambrian Rhondonia terrane of South America.

Figure 8.1 shows the first reconstruction of Rodinia, as drawn by Valentine and Moores in 1970. It isn't much as reconstructions go, showing simply a circular supercontinent bisected by a linear mountain belt running from east to west. Simple as it is, this reconstruction was a reasonable first attempt. Valentine and Moores felt that this linear mountain chain was a result of continental collision and suturing, resulting in a series of linked Precambrian mountain ranges they called the Pan-African-Baikalian system. Valentine and Moores's next image shows the breakup of Rodinia, and in this image the circular supercontinent is cut into slices like pieces of a pizza. The four chunks were, clockwise from 9 o'clock, North America, Baltica, Asia, and Gondwana.

Modern reconstructions of Rodinia (figure 8.2) place Australia and Antarctica as the missing piece along the west (present-day coordinates) coast of North America. This suggestion, first made by Charles W. Jefferson in 1978, has the advantage of satisfying a paleobiogeographic conundrum discussed below.[14] Jefferson's insight was first published as an abstract for the Geological Association of Canada and appeared in 1980 as an abstract for the twenty-sixth session of the International Geological

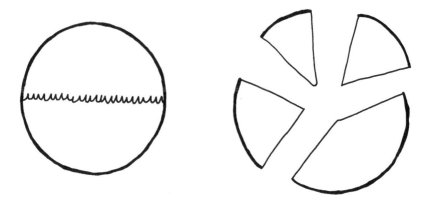

FIGURE 8.1: The first reconstruction of Rodinia, published in 1970. Intact supercontinent shown on the left, fragmentation of supercontinent into pie-shaped wedges on the right.

Congress in Paris. Jefferson argued, "North America, Australia and Antarctica were joined from more than 1,500 to 550 m.y. ago. . . . [T]he continents separated by Lower Cambrian time and Australia together with Antarctica drifted away to join [Gondwana]."

This reasoning is prophetic, although Jefferson does have North America, Australia, and Antarctica juxtaposed for a longer stretch of geological time (1500 to 550 m.y.) than is now thought to have been the case (1000 to 700 m.y.). At the time Jefferson's abstracts were published, many geologists were skeptical of Jefferson's claims, and some thought that he was badly overinterpreting his data. As the title of his abstract indicates, he was basing his continental reconstruction on lithostratigraphic correlation of the strata without other lines of evidence such as biostratigraphic correlation or paleobiogeographic similarity between North America and Australia/Antarctica. Many scientists were thus unwilling to accept Jefferson's scheme. Nevertheless, a map reconstruction of Australia and North America juxtaposed was published in 1985 by R. T. Bell and Jefferson.

Jefferson shrewdly considered the paleobiogeographic implications of his theory, however, and in an unpublished manuscript around 1980 with the same title as the 1978 and 1980 abstracts, Jefferson predicted that the Ediacara-type biota would be found near the base of the Backbone Ranges Formation in northwestern Canada. Such fossils do occur in northwestern Canada, but lower in the section, in the Blueflower and Sheepbed Formations. I would like to renew Jefferson's prediction, which has not yet been fully confirmed, that *Dickinsonia, Spriggina,* and *Tribrachidium* may

FIGURE 8.2: The modern reconstruction of Rodinia, shown as a schematic for a proposed Rodinia medallion. Various continental blocks are portrayed as follows: I. = India, ANT. = Antarctica, AUS. = Australia, N.AM. = North America, SIB. = Siberia, G. = Greenland, BAL. = Baltica.

indeed occur in the Backbone Ranges Formation, and a concerted effort should be undertaken to find them there.

I was the first to publish a paleobiogeographic analysis of the problem, "premature" as it may have been.[15] As I pointed out in 1982, distinctive members of the Ediacaran biota, including *Dickinsonia*, *Spriggina*, and *Tribrachidium* (three of the most recognizable of the Ediacaran genera), appeared in what would seem to be opposite ends of the world, namely, the White Sea region of Russia (on the continent of Baltica) and the Flinders Ranges of Australia.[16] Fedonkin had missed this in his 1983 article, assuming that the Ediacaran biota was uniformly cosmopolitan.[17] In my 1982 paper I noted the similarity between what I called the "benthic Ediacaran fauna" of Baltica and Australia and urged that this link be taken into account in any attempt to reconstruct Neoproterozoic plate positions. This posed a major problem for J. D. A. Piper's Proterozoic supercontinent reconstructions of the 1980s, which had assumed a considerable paleogeographic separation between Australia and Baltica. Bruce Runnegar[18] and Simon Conway Morris[19] agreed with Fedonkin that the forms must be cosmopolitan despite (or maybe because of) the profound geographic separation.

My paleobiogeographic suggestion forced reevaluation of the faulty reconstructions and led to a better reconstruction by Stephen K. Donovan of the supercontinent.[20] Later, other geologists[21] used new geological data to revive Jefferson's original insight, thus answering the question[22] of whether there had been an oceanic margin to western

North America since Archean time.[23] There is now a strongly emerging consensus on the main features of the Rodinia reconstruction.[24]

There were not (and are still not) occurrences of the key taxa (*Dickinsonia, Spriggina,* and *Tribrachidium*) on the land masses thought to have been between Australia and Baltica in the late Proterozoic. Thus, it appears as if I had correctly pointed out that Australia and Baltica must have been much closer to one another during the Proterozoic then they are today in order to account for the distinct similarities between the fossils. Most geological models for Rodinia before 1990 could not account for this paleobiogeographic linkage. Bruce Runnegar categorically stated in 1982 that the "present great-circle distance between these two sites is about 130°, and it is unlikely to have been less than 90° in the past."[25] In my and Dianna McMenamin's first Rodinia reconstruction, we tried to account for the biogeographic similarity by putting Baltica right up against Australia. The North America-Australo-Antarctica[26] link received confirmation in 1991.

In 1994 Guy Narbonne reported the discovery of an interesting but diminutive new Ediacaran fossil, *Windermeria,* as a possible *Dickinsonia* relative,[27] and indeed it may be, but alternatively, it could be more closely related to members of the Erniettidae. In any case, in the Rodinia reconstruction current by 1995, the distinctive Ediacaran forms of Baltica and Australia were not too distant and must have (unless there are much worse sampling errors in the recovery of these fossils than now suspected) remained close together until the demise of these rather late Ediacaran forms.

In John Rogers's 1996 review, he sees Rodinia forming about 1 billion years ago as the amalgamation of four smaller supercontinents: Ur, Arctica, Baltica, and Atlantica. This scheme is actually a descendant of one proposed in 1969 by Patrick M. Hurley and John R. Rand in which, in their figure 9, they identify two "coherent" groupings of continents, plotted on the reconstruction of Pangea as a base map.[28] Hurley and Rand's northern grouping included what Rogers now calls Arctica (plus Baltica), and their southern grouping includes what Rogers calls Atlantica plus Ur.

In Rogers's rendering, Ur (named for the German word *Ur,* "original," and what may be the world's oldest city, Ur of the Chaldees) is composed of southeastern Africa, Madagascar, most of India, and most of Antarctica. Ur is linear or *C*-shaped, an unusual shape for a supercontinent.

Arctica consists of Greenland, Siberia, and the Canadian Shield craton of North America. Baltica, as before, consists of northern Europe west of the Urals. Atlantica is composed of eastern South America and western Africa.

Rogers has Arctica and Baltica combining to form the supercontinent Nena at about 1.5 billion years ago (an acronym for *northern Europe and North America*).[29] Geologists are wonderfully adept at the generation of jargon and in the coining of new terms, but this is by no means a gratuitous exercise. In addition to providing an essential verbal shorthand, new terms, when accepted, demarcate advances in understanding.

At 700 million years ago, Rogers has Rodinia splitting into East Gondwana, West Gondwana, and Laurasia. The two halves of Gondwana come together and remain together throughout the Paleozoic.[30] The two halves of Gondwana unite at about the time of the Cambrian boundary, and by 300 million years ago, Gondwana and Laurasia have united to become Pangea (figure 8.3). This would be an example of supercontinental episodicity that has been called the Sutton Cycle.[31]

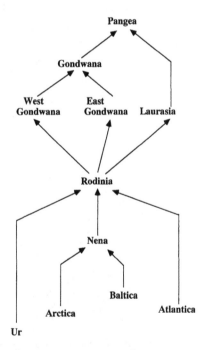

FIGURE 8.3: Diagram showing rearrangements of supercontinents over the last several billion years. Rodinia was formed approximately 1 billion years ago; Pangea formed approximately 300 million years ago.

Adapted from J. J. W. Rogers, "A History of Continents in the Past Three Billion Years," *The Journal of Geology* 104 (1996):91–107.

Rogers's analysis now allows one to construct a preliminary geological map of Rodinia (figure 8.4). The major rift fractures are present on this map, as are subsidiary rift features such as those revealed by deep seismic reflection profiling and other methods.[32]

Rodinia was formed during a 1-billion-year-old mountain-building event (orogeny) called the Grenville orogeny. The name *Grenville* is taken from the township of Grenville, Quebec, in the vicinity of the St. Lawrence River.[33] The bedrock of this region consists of marbles interstratified with gneisses, metamorphically deformed rocks that give evidence of a major Precambrian orogeny. This orogeny resulted from the fusion of Ur, Nena, and Atlantica, plus other scattered continental blocks such as East Antarctica. When continents collide in this fashion, oceanic crust is destroyed by subduction and the melted remains of the subducted oceanic slabs return to the surface as the granites and andesites of the billion-year-old Grenville orogenic belt.

From west to the east, the geologic map of Rodinia looked as follows. On the west end is Ur, consisting of cratons (continental interior blocks) in the 3-billion-year-old age range and in the 2- to 1.5-billion-year-old age range. On the east-central edge of Ur, in what appears to be a gigantic *C*-shaped embayment in Rogers's reconstruction, is a thrust belt[34] of Grenville age.[35]

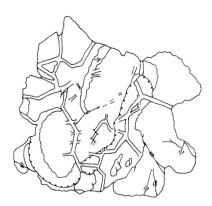

Figure 8.4: Geological map of Rodinia. Orientation roughly the same as in figure 8.2; see figure 8.2 for an explanation of the continent masses. Prominent features on this geologic map include the rift canyons (linear regions marked with double lines) and the Great Thrust Fault (marked by a single line with "teeth" on the right side). Displacements on vertical faults are marked by arrows. Note the North American midcontinent rift zone just below the center of this map.

This thrust belt has several small patches of 3-billion-year-old crustal blocks. The rocks were thrust to the west, and a dark line with teeth on the map denotes the westernmost limit of thrusting. South of this is an interior magmatic belt, also of Grenville age. Nested into the Nena embayment is most of East Antarctica, recognizable as a huge pre-Grenville craton 2.4 billion years in age. Continuing east into North America, we have here a cratonic core, the central part of the Canadian Shield, like East Antarctica a pre-Grenville craton 2.4 billion years in age. This cratonic core is surrounded by 2- to 1-billion-year-old juvenile crust, accreted around the ancient cratonic core as a number of smaller plates. Paul Hoffman has called this the "United Plates of America."[36] To the north is yet another pre-Grenville craton 2.4 billion years in age: Siberia. Siberia is composed of the roughly coeval Aldan and Anabar shields.

Like Ur, North America also has a thrust belt on its eastern margin.[37] Northwest of the northeasterly part of the limit of thrusting line is an approximately 1.5-billion-year-old orogenic belt stretching from maritime Canada into the Scandinavian peninsula. The limit of thrusting can be traced into Scandinavia as well, which is why Baltica is joined in this reconstruction with North America along the northeast coast of Greenland. This thrust fault is the largest and longest in earth history, and could be called the Great Thrust. To the east of the thrust limit line is an exterior thrust belt, just across the thrust line from the southernmost extent of the 1.5-billion-year-old orogenic belt. East of that are the paired interior magmatic belt and exterior thrust belt, respectively, of Atlantica. In the south of Atlantica is a 1.5-billion-year-old orogenic belt, associated with the Tanzania craton (2.4 billion years old) and the West Nile craton (1.5 to 2 billion years old).

As Rodinia began to break up between 1 billion and 500 million years ago, major rift systems formed between western North America and Australia/Antarctica (the Australo-American Trough) and between eastern North America and Atlantica, the nucleus of West Gondwana. Rogers has East Gondwana (Ur) heading west and Atlantica pivoting counterclockwise. I think that Ur must then go east and north in order to get Australia close enough to Baltica for them to share late (*Dickinsonia*, *Tribrachidium*) Ediacarans.[38]

In any case, the various fragments of Gondwana collide in the wake of these rifting events to form the great southern continent. These collisions led to what geologists for years have called the Pan-African orogeny. The andesites and granites formed and uplifted by these

Gondwanan collisions have been implicated in the injection of significant amounts of phosphorus into seawater, perhaps triggering the Cambrian explosion.

We are referring here to convulsive tectonic events on a global scale. The title of another of Paul Hoffman's papers (he deserves an award for creative scientific paper titles) is "Did the Breakout [from Rodinia] of Laurentia Turn Gondwanaland Inside-Out?"[39] We are surely dealing, in the rifting of Rodinia, with a profound tectonic event that alters the surface of the earth.

However, there is something unusual about the configuration of the Rodinia supercontinent. As first pointed out by Kent C. Condie of the New Mexico Institute of Mining and Technology in Socorro, there is less juvenile crust (formed of granite and related rocks) than one would expect for a supercontinent of this size.[40] Episodic isotopic ages in rocks forming juvenile crust, along with neodymium isotopic data, indicate the presence of three major pulses of continental growth in earth history: at 2.6–2.5, 2.0–1.7, and 1.3–1.0 billion years ago and some growth in the last 700 million years. However, there is a paucity of juvenile crust in the 1.3–1.0 interval, called the Grenville interval. This indicates that during the Grenvillian collisions that formed Rodinia, there were few additions (from submarine plateaus, magmatic additions from the mantle, continent-margin arcs) of juvenile crust. Condie surmises that either there is a large quantity of undocumented Rodinian juvenile crust, or 1.3–1.0 billion years ago "was not a time of extensive mantle plume activity." If, as a result of decreased plume activity, fewer submarine plateaus formed between 2.1 and 1 billion years ago, then less juvenile material could be incorporated into the newly forming Rodinia as various continental fragments collided with one another. A lesser number of submarine plateaus would have been produced, meaning that there would be a smaller volume of oceanic terranes (composed of juvenile crust) that could be accreted to an assembling Rodinia. Perhaps the Great Thrust (which would relieve orogenic stresses) also has something to do with the paucity of juvenile crust.

Condie's results should not be taken to infer that there was a paucity of collisional tectonics in the 1.3–1.0 billion years ago—quite the contrary. The Coal Creek serpentinite[41] was emplaced into the volcanic arc sediments of the Llano Uplift, Texas, implying that brittle plate collisions were important during the formation of Rodinia at 1.2 to 1 billion years ago.[42] Perhaps the brittleness of these collisions and the rarity of juvenile crust during this time are results of the same geological factors governing the collision of the continental fragments and volcanic arcs that became Rodinia.

The breakup of Rodinia follows much later. Early evidence for its fragmentation begins between 900 and 700 million years ago with sedimentary sequences of northwestern Scotland called the Stoer Group and the Sleat-Torridon Group.[43] The history of Rodinian breakup evidently extends over hundreds of millions of years.[44]

Notes

1. See p. 29 in H. W. Menard, *The Ocean of Truth* (Princeton, N.J.: Princeton University Press, 1986).

2. P. Westbroek, "The Oceans Inside Us," *The London Times Higher Education Supplement*, November 3, 1995.

3. See p. 663 in Urantia Foundation, *The Urantia Book* (Chicago: Clyde Bedell, 1955 [first written 1934]).

4. Urantia Foundation, 1955.

5. The name *Urantia* may be derived from *Urania*, the personification of astronomy.

6. Page 660.

7. Page 662.

8. See pp. 663–664.

9. J. K. Wright, "Foreword," in C. H. Hapgood, *Maps of the Ancient Sea Kings*, pp. ix–x (Philadelphia: Chilton Books, 1966).

10. J. W. Valentine and E. M. Moores, "Plate-Tectonic Regulation of Faunal Diversity and Sea Level: A Model," *Nature* 228 (1970):657–659.

11. See p. 95, M. A. S. McMenamin and D. L. S. McMenamin, *The Emergence of Animals: The Cambrian Breakthrough* (New York: Columbia University Press, 1990).

12. Page 660.

13. J. J. W. Rogers, "A History of Continents in the Past Three Billion Years," *The Journal of Geology* 104 (1996):91–107; C. Zimmer, "In Times of Ur," *Discover* 18 (1997):18–19.

14. C. W. Jefferson, "Correlation of Middle and Upper Proterozoic Strata Between Northwestern Canada and South and Central Australia," *Geological Association of Canada, Program with Abstracts* 13 (1978):429; C. W. Jefferson, "Correlation of Middle and Upper Proterozoic Strata Between Northwestern Canada and South and Central Australia," *International Geological Congress, 26th Session, Paris, Abstracts* 2 (1980):595; R. T. Bell and C. W. Jefferson. "An Hypothesis for an Australian-Canadian Connection in the Late Proterozoic and the Birth of the Pacific Ocean," in *Proceedings Pacific Rim Congress 87. An International Congress of Geology, Structure, Mineralization and Economics of the Pacific Rim. Gold Coast, Australia, 26–29 August 1987*, pp. 39–50 (Parkville, Australia: The Australasian Institute of Mining and Metallurgy, 1987).

15. See M. F. Glaessner's letter in chapter 5.

16. M. A. S. McMenamin, "A Case for Two Late Proterozoic—Earliest Cambrian Faunal Province Loci," *Geology* 10 (1982):290–292.

17. M. A. Fedonkin, "Ekologia dokembrijskikh metazoa belomorskoj bioty," in L. A. Nevessaya, ed., *Problemy ekologii fauny i flory drevnikh bassejnov*, pp. 25–33 (Moscow: Akademiya Nauk SSSR, Trudy Paleontologicheskogo Instituta, Tom 194, Izdatel'stvo "Nauka," 1983).

18. B. Runnegar, "Oxygen Requirements, Biology and Phylogenetic Significance of the Late Precambrian Worm *Dickinsonia*, and the Evolution of the Burrowing Habit," *Alcheringa* 6 (1982):223–239.

19. S. Conway Morris, "The Ediacara Biota and Early Metazoan Evolution," *Geological Magazine* 112 (1985):77–81.

20. S. K. Donovan, "The Fit of the Continents in the Late Precambrian," *Nature* 327 (1987):130–141.

21. E. M. Moores, "The Southwest U.S.-East Antarctic (SWEAT) Connection: A Hypothesis," *Geology* 19 (1991):425–428; I. Dalziel, "Pacific Margins of Laurentia and East Antarctica-Australia as a Conjugate Rift Pair: Evidence and Implications for an Eocambrian Supercontinent," *Geology* 19 (1991):598–601; M. E. Brookfield, "Neoproterozoic Laurentia-Australia Fit," *Geology* 21 (1993):683–686; I. W. Dalziel, "Earth Before Pangea," *Scientific American* 272 (1995):58–63; E. Moores, "The Story of Earth," *Earth* 5 (1996):30–33; E. M. Moores and R. J. Twiss, *Tectonics* (New York: W. H. Freeman, 1995).

22. J. P. N. Badham, "Has There Been an Oceanic Margin to Western North America Since Archean Time?" *Geology* 6 (1978):621–625.

23. P. E. Hoffman and J. P. N. Badham, "Comment and Reply on 'Has There Been an Oceanic Margin to Western North America Since Archean Time?' " *Geology* 7 (1979):226–227; W. F. Tanner, "Comment on 'Has There Been an Oceanic Margin to Western North America Since Archean Time?' " *Geology* 7 (1979):6.

24. J. J. W. Rogers, "A History of Continents in the Past Three Billion Years," *The Journal of Geology* 104 (1996):91–107; T. H. Torsvik, M. A. Smethurst, J. G. Meert, R. Van der Voo, W. S. McKerrow, M. D. Brasier, B. A. Sturt, and H. J. Walderhaug, "Continental Break-Up and Collision in the Neoproterozoic and Palaeozoic: A Tale of Baltica and Laurentia," *Earth Science Reviews* 40 (1996):229–258; I. W. D. Dalziel, "Neoproterozoic-Paleozoic Geography and Tectonics: Review, Hypothesis, Environmental Speculation," *Geological Society of America Bulletin* 109 (1997):16–42; R. Unrug, "Rodinia to Gondwana: The Geodynamic Map of Gondwana Supercontinent Assembly," *GSA Today* 7 (1997):1–6.

25. Runnegar, 1982, p. 223.

26. To which we alluded in *Emergence*, p. 99.

27. G. M. Narbonne, "New Ediacaran Fossils from the Mackenzie Mountains, Northwestern Canada," *Journal of Paleontology* 68 (1994):411–416.

28. P. M. Hurley and J. R. Rand, "Pre-Drift Continental Nuclei," *Science* 164 (1969):1229–1242.

29. C. F. Gower, A. B. Ryan, and T. Rivers, "Mid-Proterozoic Laurentia-Baltica: An Overview of Its Geological Evolution and a Summary of the Contributions Made by This Volume," in C. F. Gower, T. Rivers, and A. B. Ryan, eds., *Mid-Proterozoic Laurentia-Baltica, Geological Association of Canada Special Paper 38* (1990):1–20.

30. Unrug, 1997.

31. J. Sutton, "Long-Term Cycles in the Evolution of Continents," *Nature* 198 (1963):731–735; P. F. Hoffman, "Speculations on Laurentia's First Gigayear (2.0 to 1.0 Ga)," *Geology* 17 (1989):135–138.

32. M. A. Noe-Nygaard, "Comparaison entre les roches grenues appartenant a deux orogénies Précambriennes voisines au Groënland," *Colloque International de Pétrographie sur "Les échanges de matières au cours de la genèse des roches grenues acides et basiques," C. N. R. S., Colloque* 68 (1955):61–75; A. J. Pedreira, "Possible Evidence of a Precambrian Continental Collision in the Rio Pardo Basin of Eastern Brazil," *Geology* 7 (1979):445–448; Z. Garfunkel, "Contributions to the Geology of the Precambrian of the Elat Area," *Israel Journal of Earth Science* 29 (1980):25–40; J. A. Brewer, L. D. Brown, J. E. Steiner, J. E. Oliver, and S. Kaufman, "Proterozoic Basin in the Southern Mid-Continent of the United States Revealed by COCORP Deep Seismic Reflection Profiling," *Geology* 9 (1981):569–575; A. Piqué, "Northwestern Africa and the Avalonian Plate: Relations During Late Precambrian and Late Paleozoic Time," *Geology* 9 (1981):319–322; J. J. Peucat, P. Vidal, G. Godard, and B. Postaire, "Precambrian U-Pb Zircon Ages in Ecologites and Garnet Pyroxenites from South Brittany (France): An Old Oceanic Crust in the West European Hercynian Belt?" *Earth and Planetary Science Letters* 60 (1982):70–78; J. F. Lindsay, R. J. Korsch, and J. R. Wilford, "Timing the Breakup of a Proterozoic Supercontinent: Evidence from Australian Intracratonic Basins," *Geology* 15 (1988):1061–1064; H. J. Harrington, R. J. Korsch, and D. Wyborn, "The Rodinian Breakup Zones in the Transantarctic Mountains and Southeastern Australia," *Geological Society of Australia Abstracts* 41 (1996):183.

33. C. R. Van Hise and C. K. Leith, *Pre-Cambrian Geology of North America, U.S. Geological Survey Bulletin No. 360* (Washington, D.C.: Government Printing Office, 1909).

34. A thrust belt is a large body of rock that has been thrust over other rocks along a low-angle thrust fault.

35. Approximately 1 billion years in age.

36. P. F. Hoffman, "United Plates of America, the Birth of a Craton: Early Proterozoic Assembly and Growth of Laurentia," *Annual Review of Earth and Planetary Sciences* 16 (1988):543–603.

37. H. O'Sullivan, *Explorations Between Lake St. John and James Bay* (Quebec: Department of Colonization and Mines, 1901); O. O'Sullivan, "Explorations Along the National Transcontinental Railway from La Tuque Westward," *Geological Survey of Canada, Summary Report* (1907):67; H. C. Cooke, "Some Stratigraphic and Structural Features of the Pre-Cambrian of Northern Quebec," *Journal of Geology* 27 (1919):65–382; A. Ludman, "Significance of Transcurrent Faulting in Eastern Main and Location of the Suture Between Avalonia and North

America," *American Journal of Science* 281 (1981):463–483; G. M. Young, "The Grenville Orogenic Belt in the North Atlantic Continents," *Earth-Science Reviews* 16 (1980):277–288; M. D. Max, "Extent and Disposition of Grenville Tectonism in the Precambrian Continental Crust Adjacent to the North Atlantic," *Geology* 7 (1979):76–78.

38. For an alternative explanation, see C. Franklin, "A Biogeography of the Assembly of Gondwana," Unpublished master's thesis, Mount Holyoke College, South Hadley, Mass., 1997.

39. P. F. Hoffman, "Did the Breakout of Laurentia Turn Gondwanaland Inside-Out?" *Science* 252 (1991):1409–1412.

40. K. C. Condie, "Rodinia and the Missing 1-Ga Juvenile Crust," in *Proterozoic Evolution in the North Atlantic Realm, Program and Abstracts*, pp. 39–40 (Goose Bay, Labrador: Canadian Geoscience Council, 1996).

41. A green, snakeskin-colored rock representing a hydrothermically altered basalt, an ophiolite, or a stranded fragment of ancient seafloor crust.

42. J. R. Garrison, Jr., "Coal Creek Serpentinite, Llano Uplift, Texas: A Fragment of an Incomplete Precambrian Ophiolite," *Geology* 9 (1981):225–230.

43. A. D. Stewart, "Late Proterozoic Rifting in NW Scotland: The Genesis of the 'Torridonian,' " *Journal of the Geological Society of London* 139 (1982):413–420.

44. I. W. D. Dalziel, "Neoproterozoic-Paleozoic Geography and Tectonics: Review, Hypothesis, Environmental Speculation," *Geological Society of America Bulletin* 109 (1997):16–42.

9 · The Mexican Find: Sonora 1995

[T]he greatest successes have been for those who have accepted the heaviest risks.

—Henri Bergson

The greatest good fortune is always the least to be trusted.

—Hannibal to Scipio before the battle of Zama, 201 B.C.

What eventually became my dissertation project, a study of the Proterozoic rocks and fossils of northern Mexico, received an excellent start in 1982 with the help of Jack Stewart of the U.S. Geological Survey and his Mexican counterpart, Juan Manuel Morales-Ramirez (figure 9.1). In 1988 I received a National Science Foundation award in order to continue this field work in Sonora. But by 1994 I was feeling significant dismay about this Sonoran research project. Funds for the project had been spent (as a casino gambler might say, the chips were down), and analysis of the promising material collected in a 1990 expedition with Steve Rowland, Anne Dix, and others had failed to provide conclusive evidence of life in the Proterozoic Clemente Formation. Rowland and I were left with a handful of serially sectioned dubiofossils,[1] and virtually all we had to show for our efforts were a few published abstracts.[2] The project needed a major infusion of something, but with no easily publishable or fundable (two consecutive grant proposals were rejected) results, and with research funds in the mid-1990s becoming increasingly scarce, prospects for continuing the crucial work in Sonora had grown very dim.

Rowland and I were particularly upset with two rounds of reviews from our grant proposals to the National Science Foundation. Although some of the reviewers gave our proposal an excellent rating, the negative reviews had been particularly venomous, worse than anything I had seen before, and several of the reviewers demonstrated a complete lack of appreciation for the potential of the Sonoran section. It is best not to be too sensitive to anonymous negative comments on grant proposal reviews (a thick skin is an asset in these matters), but in this case Steve Rowland and I both felt that an injustice had been done.

FIGURE 9.1: Juan Manuel Morales-Ramirez, my Mexican geological colleague.

My research time was also being increasingly diverted to the Hypersea project, which was in the latter stages of completion and required significant research time.

So it was with mixed emotions that I received a call from the California Institute of Technology inviting me to join a research expedition to Sonora in the spring of 1995. On February 5 I received an e-mail message from Dave Evans, a graduate student in Joseph Kirschvink's paleomagnetism lab at the California Institute of Technology (Caltech) in Pasadena. Dave asked whether I remembered him from the Seattle meeting of the Geological Society of America annual last fall (I did) and said that he coveted the Ediacaran fossils that I had brought to the poster sessions. He asked me to fax him a copy of the critical topographic map of the Caborca region, with geological units shown, and with the localities indicated for two newly discovered volcanic flow rock units in the La Ciénega Formation.

Dave was interested in sampling those rocks for paleomagnetic analysis in order to complete a study begun by his adviser, Professor Joe

Kirschvink.[3] He noted that he was trying to organize a trip to Mexico on March 13–17 or perhaps a few days earlier. Dave was in the process of asking Caltech Professor Lee Silver's advice on the possibilities of taking a backhoe to the strata in order to expose fresh sediments suitable for paleomagnetic analysis as well.

Dave noted that on the trip he would be joined by Ian Dalziel and Fred Hutson at the University of Texas at Austin, who were interested in correlating the Mexican section with sections in Antarctica. Such a correlation might help test the hypothesis that Antarctica was the missing continent juxtaposed against western North America during the time of Rodinia.

I told Dave Evans that I was interested in joining the expedition. Sensing that they would be engaging field partners, I tried to contact both Dalziel and Hutson. Fred Hutson wrote me by e-mail on February 21, noting that he was interested in doing samarium-neodymium and uranium-lead isotopic work on the shales and sandstones, and carbon, oxygen, and strontium isotopic work on the limestones and dolostones in order to compare them with similar data from the Shackelton Range section in Antarctica. Noting that I had no grant funds at the time, Kirschvink and Evans generously agreed to cover half of my plane ticket to California. I replied to Dave on February 24:

> I heard from both Dalziel and Hutson, both sound like fun people in the field. Here are my flight arrangements:
>
> Arrive Burbank United flight 379, 6:00 P.M. March 12
>
> Depart Ontario airport United flight 774, 11:30 P.M. (red eye!) March 21
>
> Total ticket cost: $425, I put it on plastic. I'm glad you can cover half that.
>
> If I can stay with you at least the night of March 12 that would be good. I'm assuming camping in the field so I'll be equipped with sleeping bag, Leatherman, etc.
>
> This has the makings of an interesting expedition even if I don't find any fossils.

I heard from Ian Dalziel over e-mail on February 24. He stated that he was looking forward to the upcoming trip and invited me to join in a comment-and-reply response[4] to an article recently published by Grant Young in the journal *Geology.*

The trip was completely arranged by the end of February. Two things made this expedition possible on such short notice. First was the generous collegiality of Evans, Kirschvink, Dalziel, and Hutson, not to mention Evans's abilities as organizer. Second was e-mail, which proved to be the perfect medium of communication for arranging an expedition of widely scattered scientists.

Dave picked me up at the Burbank airport on the night of March 12. We drove first to Joe Kirschvink's laboratory at Caltech. There we had lengthy and enjoyable discussion about our science, its politics, and Joe's latest gizmo, a gun for measuring magnetic susceptibility. In addition to being a brilliant scientist and supportive colleague, Kirschvink is a world-renowned technophile with a knack for designing innovative equipment, techniques, and tools. During the discussion over the magnetic susceptibility measurements, I met Jose Hurtado, a Caltech undergraduate working in Joe's laboratory.

The magnetic susceptibility gun resembles a cross between a science fiction ray gun and a metal detector; we took turns "shooting" each other. Joe had designed it to distinguish between different types of rocks. The gun was impressive at distinguishing between a Moroccan ash deposit and a sedimentary rock, but it couldn't easily distinguish limestone of the La Ciénega Formation from a Puerto Blanco Formation basalt.

After the gun session Dave took me to his apartment in Burbank to spend the night. We had a long talk about the supercontinent Rodinia. As I rummaged around in my luggage to pull out my sleeping bag, to my surprise I pulled out a large (15 cm) but realistic rubber cockroach.[5] My wife, Dianna, has a mischievous sense of humor, and had placed it there hoping it would startle me in Mexico.

We left Caltech a little later on the morning of March 13 than I would have liked, but we had experienced some delays in finding all the equipment we needed. Caltech has a neat and highly functional storage room for field supplies, but we had to scrounge around for a couple of items. Another Caltech undergraduate was planning to join us, but she decided not to at the last minute. A piece of equipment shattered one of the windows of the '89 white Chevy Suburban as we were packing, but the glass remained in place so we decided to ignore the damage and get on the road. We felt bad about the broken rear window, but it was an accident, and as someone in the party pointed out, "Those Mexican roads *are* hard on vehicles!"

We headed east out of Pasadena, banking to the southeast along the northeastern slope of the San Bernardino Mountains. In our wide-ranging discussions during the drive, I mentioned to Dave and Jose that my main goal on this trip was to obtain specimens of Ediacarans, and in professional terms it was for me a "do-or-die" situation. I had invested a lot in this research project and now might be my last chance to deliver some really satisfying results.

We passed through San Gorgonio pass, host to an enormous orchard of wind turbines, all feeding electricity into the Pacific Gas & Electric power grid. The utility company is required by law to purchase this electricity. Passing these renewable energy sources, we followed California State Highway 86 as it stretched to the west of the Salton Sea, following a path directly above a major zone of transform faulting that will eventually split southern California in two, lengthening the Gulf of California northward and forming, at last, Isla California. I again thought back to my encounter with barnacle-shell sandbars years ago on the shores of the Salton Sea.

We reached the border at Calexico/Mexicali and waited patiently in the long line of cars at the border crossing. Jugglers seeking funds amused the motorists in line. Once over the border, we stopped for food supplies at a well-stocked grocery store (*supermercado*) in Mexicali. My companions were clearly accustomed to graduate student fare, and our acquisitions were heavy with canned goods. We knew our time in the field would be short, and we knew we couldn't afford to spend it on food preparation.

Soon we were back on Mexican Federal Route 2, headed east-southeast. Dave and Jose traded the driving. East of the Colorado River, Route 2 road becomes a narrow and nearly straight stretch of more than 100 mi before it reaches the Sonoran town of Sonoita. It supports much high-speed traffic, and much of the vehicular traffic is large tractor trailer trucks. There is little margin for error on this road because it lacks paved shoulders. It is within the free travel zone between Mexico and the United States, which means that U.S. citizens do not need passports or special permissions to travel in this part of Mexico. South of Sonoita such permission is required.

En route we picked up a case of Mexican beer (Corona) and a bottle of mescal, both nice to have for fieldwork in Mexico. One of the Caltech students had promised to introduce me to a mixed drink called the Baja Fog, made by drinking the neck out of a bottle of Corona and then refilling it with tequila or mescal. I tried to pay the liquor store owner with old Mexican coins from the 1980s, but he told me the money was worthless and demanded the newer bimetallic coinage.

We had to backtrack a few miles back into Sonoita to arrange for travel permits, for we had missed the permit office on the first pass and were sent back by guards at the point where Federal 2 passes out of the free travel zone. We told the permit agents that we were geologists hoping to collect rock samples. We finally made it out of the free zone and headed to the town of Caborca, roughly 60 mi due south of the international border.

Night had fallen, which ordinarily might have posed problems for finding the field locality, but fortunately we were equipped with good maps and Caltech's Global Positioning System (GPS), an extremely handy device that, thanks to triangulations using earth-orbiting satellites, allows the user to determine the precise latitude and longitude of his or her location. Heading south of the picturesque little town of Pitiquito, we followed the dirt road that led to our target site, the Sierra el Rajón (Cerro Rajón). Thanks to the GPS we pinpointed the eastward jeep trail heading into the Cerro Rajón. Our hope was to reoccupy a field site I had used extensively during the 1980s, when I was completing my Ph.D. work on the rocks and fossils of the area.

At 2:30 A.M. on March 14 our progress on the jeep trail was halted by a barbed wire fence with a padlocked gate. We were puzzled—the gate had never been locked in the 1980s—but because it was getting late, we decided to camp for the night and deal with the locked gate in the morning.

Dave and Jose wondered whether wire cutters should be used to cut the offending barbed wire. This was indeed a temptation, but I noted that as guests in a foreign nation (Jose is of Colombian extraction) we should be on our best behavior. This proved to be prudent advice, for we later learned that the road was booby-trapped a short distance beyond the locked gate.

As the sun rose, Dave and Jose drove into Pitiquito to attempt to find the owner of the padlock on the gate. Acutely aware of the time pressure on this expedition, I elected to stay behind and head into the desert alone. I took some food and a large canister of freshwater from the truck and sequestered it, along with my sleeping bag, under a creosote bush near the locked gate, camouflaged with a green tarp. There was no telling when Dave and Jose would return, and in fact they had a flat tire on the dirt road that led to Pitiquito.

My first task was to revisit on foot my old base camp, on the west side of the Cerro Rajón at outcroppings of the La Ciénega Formation (LCF). Here is an excerpt (p. 39, field notebook #6) from my field notes:

3/14/95: Dave Evans and Jose Hurtado went to find owner of padlock on gate. I head alone on foot east into the Sierra el Rajón.

10:55: Reached La Ciénega base camp.

10:13: Left last night's campsite at locked gate. 42 minute hike, steady pace but not fast.

[*I was concerned about how much of the limited field time was going to be consumed in what could become the daily hike, so I immediately made a measurement of how long it took.*]

La Ciénega base camp does not appear to have had visitors (human) for some time. I elect to go straight on to the LCF outcrops rather than heading south to one of the canyons.

Lunch 11:55: Small saddle. No fossils and progress has been slow.

I eventually did find some probable trace fossils in the rocks of the La Ciénega Formation, but nothing I found that day looked particularly promising. I wasn't anxious about being alone here in the field. Caution was in order, however; my companion for part of the day was a lone turkey vulture who swooped low, cocked his featherless head, and looked me over with an inquiring eye. The bird eventually flew off, having determined that I wasn't on the verge of collapse.

I followed a volcanic layer in the La Ciénega Formation eastward to a northern tributary of a larger canyon Jack Stewart and I had nicknamed Conophyton Canyon in recognition of the abundant *Conophyton* stromatolite fossils found in its rocks. I headed down and out Conophyton Canyon and turned in the direction of the base camp. It was a pleasant surprise to hear and see Ian Dalziel, Fred Hutson, and their students from the University of Texas calling to each other as they clambered over the rocks of the Tecolote Quartzite.

Ian Dalziel is a highly regarded geologist,[6] and I was very pleased to meet him in the field. I shed my field gear and shook his hand, and he introduced me to the rest of his field party. The rest of the field day we spent together looking at the unconformity separating the oldest Proterozoic sediments of the area from the older granitic rocks below. Jose and Dave joined us after dark, having spent the entire day tracking down the owner of the gate. They had located the land owner, who was happy to let us do fieldwork on the land.

The owner directed us to his gatekeeper, a Sonoran mountain man named Pedro. Pedro is a colorful character of Yaqui Indian descent. When I met him later he did not seem to be totally at ease in the company of *norteamericanos*. Pedro required more authoritative written permission from us before he was willing to open the gate. We were determined to get this permission from the authorities in Pitiquito the next day.

The morning of March 15 we drove back into Pitiquito. At the municipal office we met the helpful sheriff, several local authorities, the municipal secretary, and the president of Pitiquito, Valentina Ruiz. These local authorities were all sympathetic to our cause, and they wrote us a letter of introduction that they hoped would be of use in obtaining more definitive permission. The letter was as follows:

Pitiquito, Sonora, March 15, 1995.

To Whom it May Concern:

The person who has signed below . . . secretary of the town council, can give a good reference for Professor Mark Allan McMenamin and students Jose Miguel Hurtado and David Aspinwall Evans, residents of the United States who have expressed an interest in undertaking studies of geology in the vicinity of our municipality. I beg you give them the assistance necessary for securing from the appropriate authorities definitive permission to pursue their activities.[7]

We hoped that this letter would persuade Pedro to open the gate for us. With letter in hand we paid Pedro another visit.

North of the target campsite, but still on the west slope of the Cerro Rajón, Pedro's domicile consists of a cinder block cabin with one small window, no door, and very little floor space. The place smelled of decaying meat, and indeed there were numerous deceased rattlesnakes hanging from a string stretched across the ceiling of the cabin like so many socks hung out to dry. Also suspended from the ceiling was the partially rotted head of a young mountain lion. Pedro explained in Spanish that he killed the cat with his bare hands, and his thick callused fingers lent veracity to this claim. He also had said that he had boobytrapped our campsite access road on the other side of the fence with tire spikes.

We asked why he had gone to such lengths to keep people out. He wasn't very forthcoming in his answer, but he said something about "gringos making a mess" at the campsite recently. I knew I wasn't the culprit because I hadn't been back to the site in 5 years, and back then I was scrupulous about campsite cleanliness. You can imagine my curiosity as to the identity of the vandals!

Pedro was not satisfied with our letter, and he (not unreasonably) demanded to see something more substantive in writing, specifically, permission to do fieldwork from the authorities in Hermosillo or Mexico City. We were discouraged to hear this, but we headed dutifully back into town to see whether we could arrange for the permissions required by Pedro.

We were in the process of losing another day of fieldwork, and I was beginning to wonder whether this entire trip would turn out to be one part fiasco and one part wild-goose chase. The thought occurred to us that our efforts to secure a permit might be futile in any case because Pedro might not be willing to open the padlock even if we had a signed executive order from the president of Mexico himself. The sheriff was sympathetic but was unwilling to force the issue with Pedro.

We backtracked into Caborca and met my old friend Alfonso Salcido Reyna, a bootier who in the 1980s had been a graduate student in geology.[8] He had abandoned a career in geology, lamenting that there was no money in it.

Reyna's "Poncho" boot shop in Caborca smells richly of polished leather. Alfonso lent us his fax machine and telephone. We did make contact by phone with officials in Hermosillo, but no one was particularly helpful, and it was clear that we were not going to make much progress getting permits that day. The day was not a complete geological loss, however, for we sat in the Caborca town square, and Dave Evans gave us a brilliant exposition of the current state of knowledge of Proterozoic paleomagnetism and tectonics, in the way only a Caltech graduate student can. Nearby townspeople watched us curiously and inquired as to what in heaven's name we were talking about. As the day waned we returned to our campsite at the locked gate, determined to visit the localities on foot the next day.

I awoke before dawn on the morning of March 16 and readied my field gear for what I knew was going to be a long day. I double-checked the water level in my canteen (temperatures were warm to hot by midday), clambered over the locked gate, and headed down the jeep trail at top pedestrian speed. My goal for the day was to reach the Clemente Formation, a rock unit low in the Caborca sequence where, beginning in 1990, my fieldmates and I had found tantalizing fossil-like objects, but never anything that would convince skeptical reviewers to accept our various attempts at articles and grant proposals.

In the interest of saving time, I thought it made sense to avoid hiking all the way into the target base camp and instead make a southeastern diagonal route. This, in theory, would be the most direct path to the Clemente Formation. The route would take me over the fairly heavily vegetated surface of an alluvial fan, an area typically lacking in promise for the paleontologist. The rocks on this alluvial fan were out of stratigraphic context, jumbled together in a chaotic mixture, and covered with white, crusty desert cement called caliche. The crust would obscure any interesting fossils that might have been present in the rocks. So I hoped to pass quickly over this and save time by following the hypotenuse of the triangle rather than going east to the base camp and south along the flank of the range (an *L*-shaped trek), my usual route in the 1980s.

As I hurried along, I was careful in my choice of route because of the ubiquitous jumping cholla. This cactus (*Opuntia bigelovii*) commands healthy respect from pedestrians in the Sonoran desert. It is an erect plant, "usually with a single trunk and a close terminal group above of short lateral branches that are densely set with straw-colored spines, while those on the main trunk become quite black. The flowers are yellow to pale green, over an inch long above the ovary."[9]

Those short lateral branches tend to break off and adhere by vicious spines to passersby after only the slightest contact, hence the name *jumping cholla*. This is a mode of reproduction for the plant. The broken stems become propagules after they are torn, cut, or shaken off of their unwilling couriers. I had an intimate encounter with this plant in 1982, when I slipped on a loose slab of rock, fell sideways, and discovered a lateral branch stuck to my right ear and to the side of my head. Jack Stewart, whose field assistant I was at the time, gingerly clipped the spines with his pocket scissors until I was free. This took quite some time, and I would continue to feel spine tips in my ears for years to come.

As I quickly approached the outcrops of the Clemente Formation, I stepped into an innocuous-looking green bush. Feeling a shock of pain to my right leg, I withdrew my foot from the green bush and found a cholla branch firmly emplaced between the upper inner edge of my boot and the bottom of my pant leg. A second glance and it was clear what had happened. I had stepped on a juvenile cholla that was growing up through, and thus hidden by, the innocuous-looking green bush.

My momentum lost but with adrenaline coursing through my system, I sat down in unconsolidated fanglomerate to inspect the damage. The spines had sunk deeply into the skin just above and behind my inner right ankle, not far in an anatomical sense from where Achilles must have suffered his portentous injury. The spines had sunk in much deeper than in my 1982 incident. Clipping the spines was not an option here because friction between boot, clothing, and embedded spines would make walking difficult. These spines would have to be pulled.

Fortunately I had with me a Leatherman Tool. Manufactured in Portland, Oregon, when collapsed it looks like an all-metal pocket-knife. When deployed, however, it becomes a highly functional pair of needle-nose pliers *cum* wire cutter. In the handles of the pliers are nested an awl, a knife, a can opener, several sizes and shapes of screwdrivers, and a 20-cm ruler.

The tip of the pliers was covered in blood by the time I had liberated my ankle, but at least I was able to walk without too much agony. The Leatherman had literally saved the day; without it I would have been compelled to limp back to camp. And this painful but minor injury, rather than being an "Achilles ankle," turned out to be essential for the success that followed.

Moving forward with renewed appreciation for the cautionary phrase "haste makes waste," I carefully scanned the surface of the alluvial fan to avoid another mishap. Within minutes I noticed that the individual rocks of the rubble were no longer covered in desert crust; furthermore, they were all of the same type of rock, a greenish-tinted, blocky splitting shale. I recognized this particular lithology, for in a 1990 expedition I was coleading with Steve Rowland of the University of Nevada at Las Vegas, he and my student Anne Dix had picked up intriguing (but for many of our colleagues unconvincing) fossil-like objects in an identical type of rock.

I was still some distance from the main exposures of the Cerro Rajón, and I would not have realized that I was so close to bedrock had I not been standing exactly on this spot and had I not been paying rapt attention to what was directly underfoot. As bedrock yields to the onslaught of weathering, it decomposes to form a monolithologic rubble called regolith, the first stage in the formation of a soil. Such soil was already beginning to form on this spot.

Areas of rock rubble weathered in this way can be highly promising places to prospect for fossils: Clean rock surfaces from many different layers (but still part of the same rock unit and thus in stratigraphic context) are exposed to view. I have spent entire days at sites like this, for the changing position of the sun highlights different surfaces. A fossil invisible at 10 A.M. may become obvious by 3 P.M.

Within minutes I had convincing fossils.[10] Chance had surely favored the prepared mind in this case, for I had spent months slicing up samples of this rock collected on the 1990 expedition and I knew this to be a particularly promising facies for a fossil search. Ironically, I later learned that I was within a long stone's throw of the 1990 locality, which was located within the zone of mountainous outcrop. But this small patch of regolith was surrounded by caliche-covered fanglomerate and cacti, and was invisible from aerial photographs, maps, or even higher ground. One literally had to be right on top of it to see it, and even then one had to be watching the ground closely, for it was in the midst of an area that is characteristically unproductive of fossils.

Without the combination of the cactus and Pedro's obstinateness, I might have missed the fossils again as I had, after careful search with several different field parties, during the previous 13 years. From my notes (p. 45):

> Description of locality:
> The locality is not an actual outcrop but a ~60 × 100 meter patch of desert pavement, with desert varnish but no caliche. The mostly small float rocks are obviously derived from the same source, Unit 1 of the Clemente Fm. The patch is surrounded by clasts of mixed lithologies with caliche rims. The rock type is medium to fine gr. sand and sandy siltstone. Fine facies are greenish w/reddish tinge. This must be the westernmost "outcrop" of the Clemente Fm. in the Sierra El Rajón. Specimens will be numbered later. In addition to the presumed body fossils I found several convincing . . . trace fossils.

The Caltech students had lent me the GPS for the day, so in addition to searching for fossils I wanted to ensure that I had an accurate satellite fix for this cryptic locality. For reasons of military security there is an intentional wobble to the satellite fix, but a patient GPS surveyor can manually compensate for this wobble and thus obtain a highly accurate position.

I returned to camp at nightfall, tired but elated. Ian Dalziel remarked how it is always the first people out into field in the morning who are the last ones back in the evening. Wrapped in a sleeping bag I reflected on the good fortune of the find and wondered how I had managed to miss finding these fossils in earlier expeditions. Desert flash floods periodically rearrange the surface rocks in the area. Perhaps my regolith patch was a new exposure, recently exposed by a flash flood.

It was now Saint Patrick's Day, March 17, 1995, and I planned to take Jose and Dave to the new locality. As we headed down the jeep trail, we encountered Pedro's diabolical booby trap. The booby trap consisted of a half-inch-thick piece of wood about a foot long, with five nails driven through its long axis. The two longer nails pointed downward, and the three shorter nails pointed upward. The board was set into a rut of the jeep trail, anchored by the two longer nails. The three upward-pointing nails were intended to puncture the tires of a trespassing vehicle. A second copy of the trap was set in the opposite rut, a few yards down the road, so that the victim vehicle would potentially end up with two flat tires and be unable to ride on the spare alone. The wooden boards had been lightly sprinkled with dirt and sand, rendering the traps virtually invisible. Indeed, I had walked past or over them four times in days previous.

Although Pedro had warned us about them, Jose, Dave, and I were very upset and angry when we found the traps. Property rights are property rights, but this seemed to violate our sense of fair play. In this isolated desert region, a double flat tire could be as lethal as a land mine.

A bit further down the jeep trail we froze in our tracks as a 3-ft-long rattlesnake, disturbed from its morning slumber, appeared to leap into the air off of the jeep trail a few feet ahead of us. It then nosed toward us threateningly before sliding off to the side of the trail. I was reminded that there are worse mishaps than brushes with cholla.

I led Dave and Jose off the jeep trail and on to the diagonal shortcut to the new fossil locality. Dave was impressed when I led them directly to the largely featureless (except for the ubiquitous large *Euphorbia* cacti) patch of ground without using the GPS. Nevertheless, we took another satellite fix to ensure that we had the correct coordinates for the precious site.

From here we headed east toward the range front exposures. Soon we found an outcrop of the Clemente Oolite, useful because it gave us an unambiguous indication of our stratigraphic position. My fossil locality of yesterday was 5–10 m stratigraphically below the oolite.

We next turned south and headed toward the unconformity between the El Arpa Formation at the base of the Proterozoic sequence and the metamorphic rocks and granite below. Dave was in the lead. At some distance stratigraphically below the oolite, Dave stopped and exclaimed, with the wry grin of a field geologist who knows his earthy humor will be appreciated, "I found a nipple!" In his palm was a high-relief discoidal object with concentric and radial elements (figure 2.1). Indeed, the find resembled nothing so much as a woman's nipple. Its deep chocolate staining added an amusing touch of realism to the nipple appearance.

Joking aside, I realized that Dave had picked up a fossil that was not only better preserved than my finds of yesterday, but even further down in the stratigraphic section, and thus older. After 13 years of searching, I knew that this research program had finally hit pay dirt. I was aware that all of these fossils were occurring below the unconformity at the top of the Clemente Oolite, but the full implication of this did not dawn on me until I returned to my laboratory at Mount Holyoke College.

The next day, March 18, was devoted to collecting the paleomagnetic samples required by Dave and Jose. We needed to pack in large jugs of water to lubricate the portable field drill for paleomagnetic samples. This gas-powered piece of equipment looks just like a chain saw, except that where the saw should be is a cylindrical drill bit designed to remove plugs of rock a few centimeters in diameter. The plug is cut and, while it is still attached at its base to the parent rock body, is marked in indelible ink with careful characters to record measurements indicating its natural spatial orientation. Only then is it removed from its hole with a gentle tap of a rock hammer.

Paleomagicians, as they like to be called, sometimes feel guilty about the unsightly holes left in outcroppings of living rock by their labors. Some have taken to using the holes as tiny planters, filling them with soil and seed. Perhaps fresh *Lithops* seed (color plate 4) should be carried by desert paleomagicians as standard equipment. Personally I don't find the holes objectionable, and they do form an interesting, shady microhabitat for small animals, bacteria, and lichens.

The locked gate was a significant drawback for today's work—packing in the heavy equipment was quite a chore. We were in high spirits, though, and moved the heavy equipment at a steady pace. The sample collection was accomplished without incident, and we headed back to the vehicle. We reached the locked gate and vehicle by mid-afternoon and sat around relaxing for a time. I caught and released a horned toad that was scurrying around camp. Returning to my field notes (pp. 48–49):

> Returned to camp in afternoon, decided to break camp immediately and leave Sierra el Rajón, having accomplished all of the primary objectives of the mission. I said to Dave and Jose that there is an old Irish blessing, "May you be in heaven half an hour before the devil knows you're dead." Drove out of Pitiquito (after thanking the folks at the Municipal Office for their help in our attempts to gain permission to collect in the Rajón) and Caborca, and headed out on Rt. 2 headed west, paralleling the border. Drove on into the night.
>
> Near disastrous mishap at 10:00–10:30 P.M. Jose attempted to pass a bus but misjudged the proximity of the oncoming traffic. He went off the road to the left, pulled the vehicle back onto the road, but the vehicle was pitching and he lost control. It careened off of the road to the right. It came to a sudden stop on a level sandy area after nearly (but not) rolling over. Contents and passengers unharmed—we were very fortunate to escape serious injury. [Dave later noted that we probably would have rolled when we careened off the embankment if we hadn't been driving a fat old Chevy Suburban. Our lives may have been saved by the Mexican engineers who graded the shoulders of Route 2 with soft sand.]
>
> We were able to pull the truck back onto the highway and continue on (Dave driving). Headed west with the intent of limping across the border, getting a motel room, and having a mechanic check out the vehicle—Dave noted that the

steering was pulling to the right. At about 11:30, however, the front right tire blew out. We pulled off the highway, and then backed a safe distance off the highway to change the tire. One of the lug nuts took two people [me and Dave] to remove with violent jerk. Also, had to use cement blocks and the camp table to elevate the vehicle-the jack was too short. Finally got the spare on, and drove slowly into San Luis on the Colorado, feeling rather tired. Took a room at the Casablanca Motel (17.00 pesos—not first rate accommodations—greasy light switches and two squashed cockroaches [real ones!] on the bathroom floor). Bed springs nonexistent on one bed, but hey, it's home for the night.

3/19: Crossed border with no problems (except loss of some Corona beer to Customs) and purchased a new tire in Yuma. Installers of tire noted no damage to vehicle, but it still pulls to the right. Made it into Indio by about 1:30 P.M. Had lunch and date shakes at Oasis Date Gardens (59–111 Hwy. 111, Thermal, Cal. 92274, (619) 399–5665). This buoyed our spirits considerably. Drove through San Gorgonio pass. Very smoggy and windy, many electricity-generating windmills. Arrived Caltech late afternoon. Called my parents in Los Alamitos and Dianna and kids in Massachusetts from Dave's graduate student office. Dave met up with his girlfriend, Jean. Said goodbye to Dave, Jose, Dave's roommate Bob, Jean just as my parents picked me up in front of the Caltech seismology display for the public. Went out to dinner with Mom and Dad, was great to see them. Drove home to 3211 Oak Grove, Los Alamitos. Spent night in [my brother] Matt's old room. Parents have house up for sale, Craig Chamberlain Realtor.

Thus concluded our successful, short but eventful expedition. Back in Massachusetts I unpacked the specimens and began to carefully scrutinize them for the first time. The first step with a new find is to, if possible, compare it with previously described forms.

I have in my office a well-organized collection of articles and papers describing members of the Ediacaran biota. I use a system borrowed from the late, great Preston Cloud in which manuscripts and reprints are numbered as they come in, stored upright in boxes, and entered into an alphabetized filing system. As you might imagine, computers have vastly enhanced the functionality of such a system since Cloud's day.

With the unpacked specimens spread before me, in only a few minutes I had retrieved articles from my database with descriptions of fossils very similar to the new forms. I recognized among my fossils *Sekwia, Cyclomedusa,* a member of the Erniettidae, and two types of trace fossils that were known to co-occur with Ediacaran body fossils elsewhere. This was very exciting, for over the next few days, with a second batch of stratigraphically oriented references in hand, it was becoming clear that these new specimens, occurring as they did below the Clemente Oolite and its unconformity, were very ancient indeed.

By the end of the week I realized that I had found the world's oldest Ediacaran fossils.

Kevin McCaffrey is the news services director in the office of communications at Mount Holyoke College. He is humorous, quick-witted, and energetic, and never misses an opportunity to garner favorable publicity for the college. We communicated by e-mail shortly after my paleontological epiphany in the lab, mostly about publicizing the newly published Hypersea theory. As an aside, I mentioned to him that I had just found the oldest Ediacaran fossils, and he asked me to fire him off a one-paragraph press release. I did so, not expecting that much would come of it despite Kevin's talents.

I subsequently received a telephone call from Stan Freeman, a reporter at the local *Chicopee Union News*. Freeman was interested in the find, and I agreed to talk to him about it and allow a staff photographer to take a photograph of me and the fossils. A photographer appeared at my house, where I held the fossils for safekeeping, and took some shots of me holding the *Cyclomedusa*. "Kevin's done it again" I thought, expecting an inner page article to appear sometime in the next few weeks.

On March 31, 1995, I started receiving telephone calls and e-mail messages from friends and colleagues, telling me that I had made the front page of the paper.[11] Indeed, there was my smiling face in full color, taking up most of page 1, with my hand holding up the high-relief, concentric fossil (color plate 17).

The story caught the attention of the Associated Press, and I was interviewed by AP reporter Jeff Donne, and the story went national. David Chandler at *The Boston Globe* and Walter Sullivan at *The New York Times* wrote articles on the find. Hundreds of newspapers across the country and in Europe carried the story,[12] and the story eventually made it as far as Myanmar.

The fact that I was getting so much press for this find, and the fact that the press announcement was made before the results were scrutinized by reviewers of a professional journal, disquieted some of my colleagues. However, my policy is that it is acceptable to talk to the press before formal publication if you have important and valid results (in this case there was no doubt that the fossils were genuine) and plan to submit the results to formal review at the earliest opportunity (I was already preparing a manuscript for submission).

Some of the grumblings from my rankled colleagues were distinctly petty. My feeling is that we should get results out to the public quickly, thus helping to keep enthusiasm high for the kind of work we do. I may be the current record holder in this regard, with only 15 days having elapsed between picking up the fossils in Mexico and front-page newspaper coverage in Massachusetts. Certainly results such as these must be (and later were) formally reviewed and critiqued, but I see no reason why this process must proceed in complete secrecy.

The story went AP for a second time in October when Stan Freeman did a follow-up article announcing that the paper describing the fossils had been accepted by the *Proceedings of the National Academy of Sciences USA*.[13] The story went national again, this time appearing with photographs of the *Cyclomedusa* and another shot of me holding the fossil. Most of these stories were very well done, an exception being an article published in the *Daily Collegian* of the University of Massachusetts Amherst. The title of this was "Mt. Holyoke Prof Finds 6 [sic] Million Year Old Fossil." Oh well!

Mexicans and geographers alike might be dismayed to see the headline in the *Goshen News* of Indiana, "Super-Old Fossil Found in U.S.," or the *Albuquerque Tribune*'s announcement, "Fossil Found in Phoenix May Be Oldest Recorded." The *Albuquerque Journal* got it right: "Mexico Fossil May Set Record." From the *Arkansas Democrat-Gazette*, "Possibly Oldest Animal Fossil Is Found in Arizona"; from the *Allentown Call*, "Fossil of Jellyfishlike Creature Found in Ariz. Could Be Oldest."

A number of papers mistakenly announced that I had found the oldest fossil known (it would be the oldest animal fossil, assuming the fossils represent ancient animals). And the *Sacramento Bee* reported "Scientist on Hike May Have Found Oldest Animal Fossil," as if I were on a leisurely pleasure hike. Still others had me tripping on the specimen: "Geologist Stumbles over Ancient Find."

As the time approached for the professional article to be published, I received a telephone call from Eric Niiler of the *Patriot Ledger*, Quincy, Massachusetts. Niiler was preparing an article on Ediacarans and wanted to speak with me about the Sonoran find. He interviewed me over the telephone, and I sent him photographs of some of the specimens. Some time later he telephoned again and remarked that some of my colleagues had disparaging things to say about the validity of my results. He quoted a well-known geologist of the Boston area as saying that the discovery was "totally bogus." Although I was shocked to hear a supposedly reputable scientist say something like this before he had even seen the data or even a manuscript of the paper, it sounded like some of the things I have heard said in this competitive field. I replied to Niiler that it would appear that my colleague was not prepared "to toss any bouquets."

Shortly afterward, I received an e-mail message from the well-known geologist of the Boston area questioning Niiler's abilities as a science reporter and disavowing the "bogus" comment, adding that Niiler was trying to stir up trouble by pitting me against scientists in the Boston area. The geologist invited me to collaborate with him on radiometric dating of the Sonoran rocks, but in my reply on April 16, 1996, I politely declined his offer of help:

> [Niiler's] text as I heard it includes a number of rash statements attributed to you. Any science reporter worth his salt would not have to manufacture a dispute in order to get an interesting story from the full range of contemporary Ediacaran research.
>
> I won't be needing help with the geochronology of the section, but thank you for your kind offer of assistance. I will rush you a copy of the PNAS paper as soon as it is available.

After Kevin McCaffrey and I spoke further to Niiler, trying to impress upon him that his reputation as a science writer was in jeopardy if he botched this article, he agreed to omit the more objectionable rhetoric.

A pair of articles by Niiler appeared in the April 23 issue of *The Patriot Ledger*, one titled "Creatures of Creation: Fossil Discoveries Shed New Light on Evolution of the First Animals" and a second, "Mass. Geologist Creates Paleontological Debate." A large photograph of the well-known geologist of the Boston area accompanied the articles, along with the photograph of the Sonoran *Cyclomedusa*. In the latter he is cited (p. 19) as saying, "He [McMenamin] has no idea how old those rocks are" but followed with the disclaimer that said scientist has not "reviewed McMenamin's scientific paper." In spite of these prickly comments, Niiler's articles were not badly done.[14]

This episode was a lesson for me in publicity hardball. With regard to the outcome, I felt that I had fought it to a draw this time. As a colleague once said to me, "There is no such thing as bad press as long as they spell your name right."

The published paper finally appeared on May 14, 1996, in the *Proceedings of the National Academy of Sciences (USA)*.[15] This article describes the fossils and places them in their paleobiological and stratigraphic context. It criticizes geologists in the Boston area for being "incautious" in their published assumptions that the Ediacarans were animals. My paper also questions the validity of supposed animal fossils found in the Twitya Formation of Canada.[16] These putative fossils occur below glacial deposits and if confirmed would be older than the Mexican fossils. The Twitya structures are unconvincing as fossils (and could easily be pseudofossils), however, and furthermore are not associated with trace fossils, as are the Mexican discoveries. Canadian geologists must restudy the Twitya Formation and locate some genuine fossils if they wish to overturn my record.

Shortly after publication, I received a call from Richard Monastersky, the earth science reporter for *Science News.* He interviewed me for an article on the find and asked whether I had any color photographs of the specimens.

Monastersky's article, "Living Large on the Precambrian Planet," appeared as a *Science News of the Week* feature article.[17] The Mexican *Cyclomedusa* appears in full color, and Monastersky's text describes how paleontologists can now identify both the oldest and the youngest Ediacarans. Peter Crimes and his co-workers at the University of Liverpool, in a 1995 article in *Geological Journal,* described Ediacaran specimens in Upper Cambrian (520- to 510-million-year-old) rocks of County Wexford, Ireland.[18] The Irish specimens (figure 2.2) consist of discs with both concentric and radial elements, and are in some features very reminiscent of the largest Mexican fossil. Crimes, in a challenge to Seilacher,[19] noted that "there was no mass extinction at the end of the Precambrian."

Monastersky emphasized the ages of the new finds, fairly reported on the controversy surrounding the age of the Mexican material, and concluded that although the Mexican and Irish finds broaden the range of the Ediacaran biota, they do not silence the debate about these creatures. He cites Smithsonian paleontologist Douglas H. Erwin: "We now know when they lived, we just don't know what they were."

Press attention tends to generate more press attention. The cover story of the March 1997 issue of the magazine *Discover* is a splendid article on the Ediacarans by Karen Wright.[20] *Dickinsonia* graces the cover. Wright states on the contributors' page that "most science is about answers, insight, and conclusions. . . . The Ediacarans intrigued me because their story is nothing but questions."[21] She begins the article by comparing the history of life to a movie ("a period of asteroid bombardment ensures great FX!"[22]) and notes that the story of the Ediacarans has "all the elements Hollywood abhors: ambiguity, contradiction, messy subplots, unresolved endings." The article describes Wright's visit to the Mistaken Point, Newfoundland, locality: "You don't want to step on [the fossils], but you can't avoid it; they're as densely sown as wildflowers flattened in a hailstorm." Wright then goes on to address the Vendobiont controversy, the Garden of Ediacara, and question of preservation of these fossils in rocks. The Mexican discovery of 1995 gets a mention, and I'm quoted saying, "The Ediacaran biota is something of a professional embarrassment for paleontology. It's the most dramatic moment in the history of life, and we can't even [identify] the cast of characters."

By the end of March 1997 the fortunes of my stalled Sonoran research project had taken a dramatic turn for the better. I now had body fossils low in the section and was able to claim forcefully that they were the world's oldest. I have in preparation a series of new papers further describing the Clemente Formation material. I was winning the battle of the popular press.

Critics who would dispute my age claim now have to shoulder the burden of proof by finding demonstrably older Ediacaran fossils. Perhaps they will be able to find them, perhaps they will not. Either way, it is a win-win situation for the advancement of the science of geology. I propose an inviolable code of conduct for all scientists (McMenamin's Rule): Always make scientific rivalries work to the benefit of the science.

Perhaps I have overly focused here on the negative interactions with colleagues, but there is no sense in pretending that such competition does not exist, and in any case it is all for the best if we focus on the research and obey McMenamin's Rule. Adherence to this rule is one measure of scientific greatness. One of the nicest things about the Mexican discovery is how it completely overturns the negative reviews of the Rowland/McMenamin proposal from the National Science Foundation. Steve Rowland and I deserve to gloat a bit—we've been completely vindicated.

Furthermore, I could claim what I felt was a decisive victory in the battle for media attention. Hundreds of newspapers across the United States, Europe, and the world had reported the Mexican discoveries (although none, so far as I am aware, in Mexico[23]), demonstrating broad-based interest in this type of research. The *Science News* and *Discover* pieces demonstrated that although there were still questions regarding the absolute dating of the Mexican fossils, the fossil find itself and the relative age assessment attached to it were commanding respect from my colleagues.

The landscape of scientific reputations is never static, however, and it is not always easy to anticipate the next challenge. One thing was clear, however: My next task was to translate media attention into grant funds for further research in Mexico. I am confident that there are many more important fossils to be found in Sonora, and that the next phase of global Ediacaran research must focus on these Mexican sections.

Scientific rivalries sometimes seem to get as heated as the rivalries usually associated with wars. However, casualties in this case were limited to bruised egos and injuries caused by a few cactus spines. Perhaps the best thing about paleontology is that the rivals depend on one another for new data. More important, when my colleagues and I are finished shooting off our mouths and pens, the smoke clears, and we can all walk away. That is, assuming we haven't had too many Baja Fogs!

Notes

1. Dubiofossils are structures that look like fossils but are not convincingly biogenic.

2. A. Dix and M. McMenamin, "A New Fossil from the Vendian-Cambrian Boundary," *Abstracts of the Sixteenth Annual Undergraduate Science Symposium, Mount Holyoke College* 16 (1991):16; M. A. S. McMenamin, S. M. Rowland, F. Corsetti, A. M. Dix, and R. P. Nance, "Vendian Body Fossils (?) and Isotope Stratigraphy from the Caborca Area, Sonora, Mexico," *North American Paleontological Convention Abstracts* 6 (1992):206; M. A. S. McMenamin, S. M. Rowland, R. P. Nance, and F. Corsetti, "Proterozoic Fossils from Mexico," *29th International Geological Congress Abstracts, Kyoto, Japan* 2 (1992):257; S. M. Rowland, F. Corsetti, and M. A. S. McMenamin, "Carbon Isotope Stratigraphy of the Proterozoic-Cambrian Section of the Caborca Area, Sonora, Mexico: Preliminary Results," *Geological Society of America Abstracts with Program* 25, no. 5 (1993):140.

3. T. D. Barr and J. L. Kirschvink, "The Paleoposition of North America in the Early Paleozoic: New Data from the Caborca Sequence in Sonora, Mexico," *EOS* 64, no. 45 (1983):689–690.

4. I. W. D. Dalziel and M. A. S. McMenamin, "Are Neoproterozoic Glacial Deposits Reserved on the Margins of Laurentia Related to the Fragmentation of Two Supercontinents?: Comment and Reply," *Geology* 23 (1995):959–960.

5. © 1987 Accouterments Seattle Hong Kong printed on its abdomen.

6. When I later invited him to speak at Mount Holyoke College, the auditorium was packed.

7. Translation by Mark McMenamin, February 25, 1997.

8. A. S. Cevallos Ferriz, A. Salcido Reyna, and A. Pelayo Ledesma, "El registro fosil del Precambrico—Los estromatolitos de Caborca, Son.," *Notas Geologicas* 2 (1982):2–6.

9. See p. 39 of P. A. Munz, *California Desert Wildflowers* (Berkeley: University of California Press, 1962).

10. M. A. S. McMenamin, "Ediacaran Biota from Sonora, Mexico," *Proceedings of the National Academy of Sciences (USA)* 93 (1996):4990–4993.

11. S. Freeman, "Local Geologist Makes Ancient Find: Fossils May Be Earth's Oldest," *Springfield (Mass.) Union-News,* March 31, 1995:1–6.

12. The following (partial list) newspapers carried an account of the story in 1995 or 1996: *ABC* (Spain), *Albuquerque Tribune, Argus Leader* (Sioux Falls, S.D.), *Arizona Daily Star* (Tucson), *Arizona Republic* (Phoenix), *Arkansas Democrat-Gazette* (Little Rock), *Banner* (Cleveland, Tenn.), *Berkshire Eagle* (Pittsfield, Mass.), *Call* (Allentown, Pa.), *Camera* (Boulder), *Cape Cod Times* (Hyannis, Mass.), *Chicopee-Holyoke Union-News, Courier News* (Blytheville, Ark.), *Crescent-News* (Defiance, Ohio), *Daily Collegian* (University of Mass., Amherst), *Daily Hampshire Gazette, Denver Post, Dispatch* (Brainerd, Minn.), *Dispatch* (Moline, Ill.), *Dispatch* (Oneida, N.Y.), *Eagle/Times* (Reading, Pa.), *Enquirer* (Battle Creek, Mich.), *Flint Michigan Journal, Gazette* (Billings, Mont.), *Gazette* (Phoenix), *Gazette Telegraph* (Colorado Springs), *Gazette* (Kalamazoo, Mich.), *Globe* (Dodge City, Kans.), *Hackensack (NJ) Record, Hartford Courant* (Conn.), *Herald* (Brownsville, Tex.), *Herald* (Clinton, Iowa), *Herald* (Sanford, N.C.), *Herald-Leader* (Lexington, Ky.), *Herald-Post* (El Paso, Tex.), *Idaho Press-Tribune* (Nampa-Caldwell), *Idaho Statesman* (Boise), *Indianapolis Star, Intelligencer* (Wheeling, W.V.), *Journal* (Albuquerque), *Journal* (Kankakee, Ill.), *Journal* (Stevens Point, Wis.), *Journal Tribune* (Biddeford-Saco, Maine), *Kansas City (Mo.) Star, Los Angeles Times, Lynn MA Item, Morning News* (Blackfoot, Idaho), *Morning Star Telegram, Mountain Press* (Sevierville, Tenn.), *New York Times, Newburyport (Mass.) News, News* (Amarillo, Tex.), *News* (Andarko, Okla.), *News* (Bowling Green, Ky.), *News* (Hutchinson, Kans.), *News* (Jacksonville, N.C.), *News* (Newport, R.I.), *News* (Port Arthur, Tex.), *News* (Salem, Mass.), *News and Observer* (Raleigh, N.C.), *News-Gazette* (Champaign-Urbana, Ill.), *News-Journal* (Daytona Beach, Fla.), *News-Leader* (Springfield, Mo.), *News-Post-Herald* (Birmingham, Ala.), *News-Record* (Gillette, Wyo.), *News-Record* (Harrisburg, Va.), *North Jersey Herald-News* (Passaic, N.J.), *Oregonian* (Portland), *Orlando Sentinel, Patriot* (Harrisburg, Pa.), *Patriot Ledger* (Quincy, Mass.), *Press and Sun-Bulletin* (Binghamton, N.Y.), *Press Herald* (Portland, Maine), *Press Republican* (Plattsburgh, N.Y.), *Record* (Troy, N.Y.), *Reformer* (Brattleboro, Vt.), *Register* (New Haven), *Reporter-Herald* (Loveland, Colo.), *Review-Journal-Sun* (Las Vegas), *Rocky Mountain News* (Denver), *Sacramento Bee, Sentinel* (Carlisle, Pa.), *Spectrum Northern Edition* (St. George, Utah), *Standard-Observer* (Pittsburgh), *Star* (Hope, Ark.), *Star-News* (Pasadena), *Summit Daily News* (Colorado), *Sun* (San Bernardino, Calif.), *Sun* (Yuma, Ariz.), *Sun Chronicle* (Attleboro, Mass.), *Tampa (Fla.) Tribune, Telegram and Gazette* (Worcester, Mass.), *Telegraph* (N. Platte, Nebr.), *The Boston Globe, The Philadelphia Inquirer, Times* (St. Petersburg, Fla.), *Tri-City Herald* (Pesco, Wash.), *Tribune* (Albuquerque), *Tribune* (Bismarck, N.D.), *Tribune* (Great Falls, Mont.), *U.S. News* (Goshen, Ind.), *Vindicator* (Youngstown, Ohio), *Washington Times,* and *World-Herald Morning Edition* (Omaha, Nebr.).

13. S. Freeman, "Experts Confirm Age of Geologist's Fossils: The Local Teacher's Paper on the Discovery Has Been Accepted for Publication in the *Proceedings of the National Academy of Sciences," Springfield (Mass.) Union News,* October 19, 1995:A1–A11.

14. E. Niiler, "Creatures of Creation: Fossil Discoveries Shed New Light on the Evolution of the First Animals" and "Mass. Geologist Creates Paleontological Debate," *The Patriot Ledger* 160, no. 91 (1996):18–19.

15. M. A. S. McMenamin, "Ediacaran Biota from Sonora, Mexico," *Proceedings of the National Academy of Sciences (USA)* 93 (1996):4990–4993.

16. H. J. Hofmann, G. M. Narbonne, and J. D. Aitken, "Ediacaran Remains from Intertillite Beds in Northwestern Canada," *Geology* 18 (1990):1999–1202.

17. R. Monastersky, "Living Large on the Precambrian Planet," *Science News* 149 (1996):308.

18. T. P. Crimes, A. Insole, and B. P. J. Williams, "A Rigid-Bodied Ediacaran Biota from Upper Cambrian Strata in Co. Wexford, Eire," *Geological Journal* 30 (1995):89–109.

19. A. Seilacher, "Late Precambrian and Early Cambrian Metazoa: Preservational or Real Extinctions?" in H. D. Holland and A. F. Trendall, eds., *Patterns of Change in Earth Evolution*, pp. 159–168 (Berlin: Springer-Verlag, 1984).

20. K. Wright, "When Life Was Odd," *Discover* 18 (1997):52–61.

21. Page 10.

22. Shorthand for "effects."

23. I remedied this situation myself by writing an article in Spanish on the find (M. McMenamin and H. D'Ambrosio, "La Biota Ediacara de Sonora," *Geología del Noroeste* 2 (1997):15–16.

PODER EJECUTIVO FEDERAL

CONSEJO DE RECURSOS MINERALES

NUM. 2

La presente acredita al C. MARK ALLAN McMENAMIN

VALIDA HASTA

Cuyo retrato y firma obran al calce, como INVESTIGADOR CIENTIFICO ESPE-CIAL

1982

Las autoridades Civiles y Militares deberán prestar la ayuda y garantías necesarias para el desempeño de su comisión.

México, D. F., a 5 de ENERO de 198 2.

El Director General

ING. GUILLERMO P. SALAS

FIRMA

PLATE 1 Official badge issued by the Mexican Bureau of Mineral Resources to Mark McMenamin for the conduct of government fieldwork in Sonora, Mexico.

PLATE 2 "Double-strike" mudcrack impressions from the Fish River Canyon, Namibia.

PLATE 3 Impression and counterimpression of the Ediacaran frond *Charnia masoni*. Note the zig-zag medial suture running down the center of the frond. From the Midlands region of Great Britain. Length of frond 19.3 cm.

Thompson & Morgan

LIVING STONES
Lithops

$3.99
181 925

Net Wt 4 mg

2546   GREENHOUSE PERENNIAL

PLATE 4 Seed packet cover for the "living stone" *Lithops,* sold as a greenhouse perennial.

PLATE 5 Quiver tree or kokerboom (*Aloe dichotoma*). A. Seilacher standing next to tree. Note polygonal bark.

PLATE 6 The Bahnhof Hotel in Aus, Namibia.

PLATE 7 The author at the *Pteridinium* locality, sitting on a rectangular stone structure built by German colonial *Schutztruppe* for their horses.

PLATE 8 A specimen of *Pteridinium* in which the organism has been folded back over itself. The nose of the fold is at the left-hand edge of the specimen. Longest dimension of rock in view, 5.5 cm.

PLATE 9 The port town of Lüderitz, Namibia.

PLATE 10 A *Pteridinium* specimen from the Erni collection. Both "bathtubs" are swollen with sand. Scale in centimeters.

PLATE 11 A *Pteridinium* specimen from the Erni collection with a very straight medial suture. Scale in centimeters and inches.

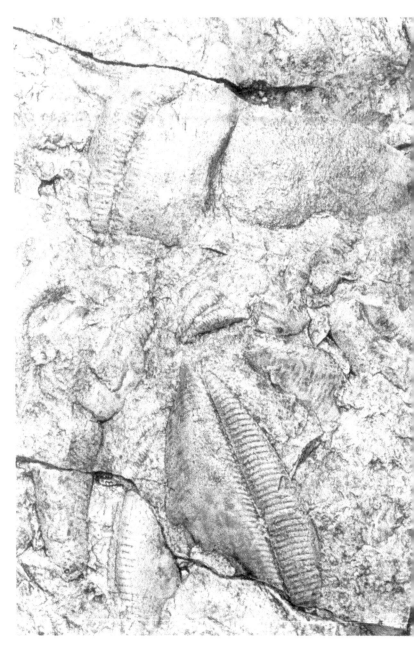

PLATE 12 An epoxy cast of the Namibian *Pteridinium* slab excavated by Seilacher's team in 1993, painted and stained by the author to simulate the color of weathered quartzite

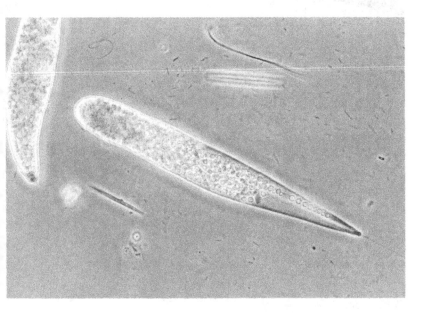

PLATE 13 *Leptoseris fragilis*, a photosymbiotic coral photosynthesizing at 120m water depth in the Red Sea. Photograph taken in the submersible *Geo* by D. H. Fricke. Photograph courtesy D. Schlichter.

PLATE 14 Individual *Ophrydium* cell. Length of cell approximately 0.7 mm. Photomicrograph by L. Margulis.

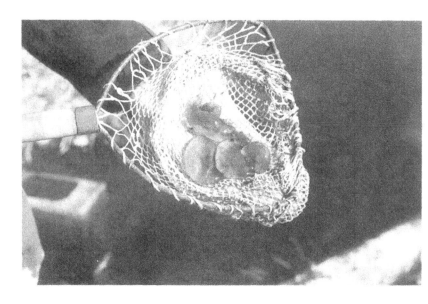

PLATE 15 *Ophrydium* colonies (up to 7 cm thick) collected at Hawley Bog, Massachusetts. Photograph courtesy T. H. Teal and L. Margulis.

PLATE 16 Apparatus designed to measure the transmission of light through wet and dry sand. See text for description.

ONCE UPON A TIME — Geology professor Mark McMenamin holds an ancient jellyfish-like fossil he found earlier this month in Mexico's Sonoran Desert. It is believed to be the oldest organism yet discovered anywhere.

Local geologist makes ancient find

By STAN FREEMAN

Fossils may be Earth's oldest

SOUTH HADLEY — What may be the oldest animal fossils ever found — jellyfish-like creatures that lived perhaps 590 million years ago — have been discovered in northern Mexico by a Mount Holyoke College geologist.

The two-inch-long impressions in shale were found two weeks ago in the Sonoran Desert by a team of geologists and paleontologists led by Professor Mark McMenamin. He decided to revisit the cactus and rattlesnake-filled terrain on a hunch, first made during an expedition there in 1980, that the rocks might contain particularly ancient fossils.

"The specimens we found in Mexico are the oldest complex organisms of any type. Previously, the oldest fossil was 575 million years," he said yesterday.

The fossils are of marine life-forms called Ediacaran biota by paleontologists.

"They are representative of what many believe to be the earliest animal life. It would be closest to jellyfish or sea anemones today, although there is a controversy as to whether they are really closely related," McMenamin said.

The fossils were found during an expedition carried out March 14-19. The team, organized and led by McMenamin, included six other people — two students from the California Institute of Technology, two faculty members and two students from the University of Texas a

Continued on Page 8

PLATE 17 The author's fossil discovery and smiling face received front-page coverage in a local Massachusetts newspaper on March 31, 1995.

PLATE 18 An unconformity in the Fish River Canyon, Namibia. The flat-lying, plateau forming rocks in the center of the photograph overlay tilted and eroded strata that di down to the banks of the Fish River. The unconformity is a roughly horizontal plane tha separates the horizontally bedded strata above from the tilted strata below.

10 · The Lost World

If Karl Popper taught scientists one thing it was that invention without testing is not science, and the reader will look in vain for ways in which the Garden of Ediacara and its fall can be critically tested: each new fact offers only support.

—R. A. Fortey[1]

Too slavish an adherence to the doctrine of falsifiability and testability can blind one to the more intuitive (and often ultimately more successful) approaches to scientific work.

—*Hypersea*[2]

An unconformity (color plate 18) is a point in a layered sequence of rocks where the direction of the layering abruptly changes direction. The transition zone is usually very sharp, and although it can be a nearly planar surface, it is more often an irregular pitted surface covered by irregularly strewn rock rubble (now turned to a rock called conglomerate).

An unconformity in Scotland gave Scottish geologist James Hutton the insight for which he is best known. He was the first to understand the mind-boggling immensity of geologic time. Hutton was so impressed by this, and by how much time was missing[3] in the unconformity zone separating the older, tilted strata from the younger, more horizontal strata deposited upon them, that he tried to argue that history, in geology, has no meaning. By arguing that the full measure of geological time is an unending cycle of uplift (tilting), erosion, and deposition, only to cycle over anew, Hutton effectively denied the earth a history. The famous last line of his book states that the earth has "no vestige of a beginning, no prospect of an end."

The great Russian geochemist Vernadsky was deeply influenced by Huttonian thought in these matters, to the point where he could not even conceive of a planet Earth without life. In other words, life is one of its most important geochemical forces. In the 1920s Vernadsky argued that any research program into the origins of life belonged in the metaphysics department, not in the sciences.

Charles Darwin, also profoundly influenced by Huttonian thought through his mentor, Charles Lyell, was troubled by the apparent lack of ancestors for the animals of the Cambrian fauna. This seemed to represent an anomalous historical time marker, an annoying vestige of a beginning. True to his preferences for gradual evolutionary change, Darwin reasoned that there was a gap in our knowledge of Cambrian ancestors, possibly caused by a gap in the sedimentary record.

Charles D. Walcott, discoverer of many important Cambrian fossils, including the famed Burgess Shale, followed Darwin's suggestion and proposed that the sudden appearance of Cambrian fossils was the result of an unprecedented break in the recorded history of life. Writing in 1914, he named this gap in the record the Lipalian interval (from the Greek word for "lost"). The Lipalian interval became

the era of unknown marine sedimentation between the adjustment of pelagic life to littoral conditions and the appearance of the Lower Cambrian fauna. It represents the period between the formation of the Algonkian continents [the fragments of Rodinia] and the earliest encroachment of the Lower Cambrian sea.[4]

Walcott felt that the sudden appearance of the Cambrian phyla resulted from the fact that the marine sediments deposited during the early stages of animal evolution were not exposed to observation by geologists, and that the unfossiliferous strata coming right before the Cambrian were mostly sediments deposited on land, sedimentary rocks that could not be expected to bear marine fossils.

The Lipalian concept was an attractive one in its day, a mysterious interval of missing sediments, a lost world during which the most important stages of animal evolution had occurred, and was a considerable source of comfort to those who were bothered by the sudden appearance of Cambrian animals. As late as 1958 geologists were still trying to support and embellish the Lipalian theory. Daniel I. Axelrod argued that the earliest animals lived exclusively in shoreline habitats (the coastline hypothesis) and were absent from the fossil record because "those deposits have largely been eroded and the records have been lost."[5] Axelrod wrote that "clearly, a significant but unrecorded chapter in the history of life is missing from the rocks of Precambrian time."

However, the Lipalian interval posed a problem for stratigraphers, whose task it was to organize the strata into coherent sequences based on the relative ages of the various layers. For example, the Jurassic system, or all the rocks deposited during the Jurassic period, are younger than all of the rocks of the Triassic system and older than all of the rocks of the Cretaceous system. How could one define a Lipalian system when the sedimentary rocks deposited during this proposed period were either missing or not exposed to view?

Shortly after the end of World War II, it became increasingly clear that the Lipalian concept was in need of modification. Professor F. G. Snyder of the University of Tennessee, in a lecture given for the Tennessee Academy of Science, Nashville, in 1946, presented what in retrospect are brilliant insights into the nature of the Lipalian problem. First he made a clear distinction between the Lipalian interval and the Precambrian unconformity. The two are not the same, for as Snyder pointed out: "Walcott specifically stated that the long Lipalian interval is 'represented by deposition of the great series of pre-Cambrian sedimentary rocks on the North American continent.' "[6]

Walcott discussed the unconformity in the same paper and claimed that "the pre-Cambrian unconformity is universal in all known localities of Cambrian sedimentation." On one hand, Walcott considered the great accumulation of North American Lipalian sedimentary rocks, and on the other, he alluded to the supposedly universal unconformity.

Snyder, making specific reference to Cambrian and Precambrian strata of western North America, obliterates the concept of the universal unconformity by stating that "contrary to the traditionally accepted view, the unconformity separating the pre-Cambrian from formations now classed as Lower Cambrian is in places neither marked nor profound,"[7] and "in some areas pre-Cambrian sediments pass into 'Lower' Cambrian sediments with little or no break in deposition."[8] Snyder's inferences have been completely borne out by subsequent work.

But the most astute observation in Snyder's paper is his interpretation of the duration of the Cambrian explosion. In a single paragraph, Snyder set the stage[9] for the next 50 years of research in paleontology: "The writer believes that evidence presented by the fossils themselves, both in the Lower Cambrian and in later periods, suggests that the period of development of shells was *far shorter* than is generally believed and was represented by continuous marine deposition" (italics mine).

Dianna McMenamin and I, in our 1990 book *The Emergence of Animals: The Cambrian Breakthrough*, arrived at the same conclusion, noting that "the apparent suddenness of the event (at most a few million years) is real."[10] Seilacher also had recognized this in his 1956 paper on trace fossils of the Cambrian and before.[11] Because of the explosion in types and numbers of both trace and body fossils, Seilacher calls the rocks of this time by names that suggest a Vernadskian twist; Seilacher characterizes the transition in trace fossils as a sudden changeover from physical, lithological dating of rocks ("Petrogaeicum") to biological dating of rocks ("Biogaeicum"). Recent geochronology research confirms the brevity of the Cambrian explosion.[12]

Snyder was apparently the first person to read the Cambrian record correctly, seeing clearly for the first time that the Cambrian event could and should be read as a sudden and major evolutionary event.[13] By stating that "so firmly entrenched is the belief in the slowness of evolution that many writers consider the Lipalian period to have extended far back into the pre-Cambrian," Snyder neatly anticipates the emergence, a quarter century later, of punctuated equilibrium theory in paleontology.[14]

Richard A. Fortey once characterized Dianna's and my reading of the Cambrian explosion as sudden as being "of course, 'punctuated equilibria' writ large,"[15] when in fact, if one considers the historical progression of thought,[16] punctuated equilibrium theory is the Cambrian explosion writ small.

Although Snyder does not speculate on the reasons for the cause of the Cambrian explosion, he does note that "organisms did not possess the ability to form shells at the beginning of Cambrian time but quickly acquired the ability when environmental conditions or pressure of competition with other forms of life necessitated rapid change." This discussion is completely in accord with the conclusions of *Emergence of Animals*, in which we attribute the appearance of skeletons to ecological changes associated with the first large predators. Snyder's paper, so far ahead of its time, clearly demonstrates how a well-constructed geological review article can still ring like boulder of Schwartzrand Limestone 50 years after its publication date.

In 1946, the same year as Snyder delivered his epochal paper, an article was published by L. Lungerhausen titled "On Certain Peculiar Features of the Ancient Series of the Western Slope of the South Urals."[17] The Urals have long been of interest to geologists because they have what was once known as a "primitive band" of fossiliferous strata flanking a granitic axis.[18] Lungerhausen's paper discussed the sedimentologic features characterizing an unfossiliferous series of shales, arkoses (feldspar-rich sandstones), and tillites (glacial deposits) occurring in the southern Urals. In sum total, these represented a massive accumulation of sediment 15 km thick. Curiously, Lungerhausen described this thick pile of ancient sedimentary rock as belonging to the Lipalian system. Like Snyder, he must have undertaken a careful reading of Walcott. Lungerhausen took it a step further, however, and proposed the Uralian sequence as the type occurrence of the Lipalian system.

It is important to consider for a moment the implications of defining a new geological system. What Lungerhausen had done, by erecting the Lipalian system, was by implication to define a new geological period, the Lipalian period, equivalent to other periods such as the Permian or Triassic period. Although Walcott strongly alluded to this, no one before had done this for Precambrian rocks, although several attempts have been made since.

The earliest of these attempts was the definition of the Vendian system and Vendian period by Sokolov in 1952.[19] This proposal was not met with universal acclaim, largely because the strata on which the Vendian system was based were known only from core samples recovered from depth.

Geologists have a fondness for being able to walk over and visually inspect important rock sequences. Even some of Sokolov's Russian colleagues rejected the concept of a Vendian system. L. J. Salop, in his influential book *Precambrian of the Northern Hemisphere*, avoided use of the term *Vendian*, noting that it was applied to deposits of too great a vertical range.[20] Preston Cloud and Martin Glaessner discounted the Vendian system because the stratotype section for the Vendian is "inaccessible to direct observation."[21]

Salop opted instead for the term *Eocambrian* (the dawn of the Cambrian). The Eocambrian was proposed in 1900 by W. C. Brögger as a global system with, as type area, the sub-Cambrian Sparagmite series of Norway.[22] A related concept, the Infracambrian, was proposed in 1949 by N. Menchikoff with, as type section, the 3050 m of limestones occurring below beds with Lower Cambrian fossils in Morocco.[23] However, these terms have never been accepted as formal systems or periods and long ago fell out of favor with Western geologists, as chronicled by Robert B. Neuman and Allison R. "Pete" Palmer.[24]

Cloud and Glaessner, following up a suggestion by Termier and Termier, offered a counterproposal, suggesting that the interval before the Cambrian be called the Ediacarian period and Ediacarian system (note the additional *i* in this spelling), with a type section in the Flinders Ranges of South Australia.[25]

Dianna and I, in *Emergence of Animals*, felt that the Ediacarian system could not be used because the Vendian system had priority. The law of priority is an important one in science: The first person to name a species, taxon, or geological entity obliges other scientists to use said name. We encouraged use of the term *Vendian*, arguing that, "It makes good sense to find a Vendian stratotype that is more accessible than the original borehole sequences, but we don't think the term needs to be replaced with the unwieldy 'Ediacarian' " (p. 88).

Members of the committee commissioned to decide such things tell me that priority does not play a role in the establishment of new geological periods. To this I say hogwash: We should honor such priority whenever we can do so, in order to maintain continuity with the work of those who came before us.

I now propose to end the confusion about what to call the strata immediately before the Cambrian and provide a geological period for the important time interval during which these rocks were deposited. I propose that, to honor the work of Charles Walcott, we use the terms Lipalian system and Lipalian period.[26]

There is considerable precedent for such a proposal. Walcott, as we saw from Snyder's bellwether paper and Lungerhausen's stratigraphy, had in mind a sequence of North American sedimentary rocks called the Lipalian. Lungerhausen first proposed a Lipalian system, but his proposal suffered from the fact that his proposed type sequence is not in North America, as Walcott would have intended, and is overlain by sediments of the Asha Series, now known from the work of Yu. R. Bekker to contain an important suite of Ediacaran fossils,[27] thus rendering Lungerhausen's rocks too early to be properly Lipalian. It seems reasonable to me that the period before the Cambrian should include the early Ediacarans.

The 1976 edition of the widely read *Dictionary of Geological Terms*, published by the American Geological Institute, defined Lipalian as follows: "A theoretical geologic period immediately antedating the Cambrian. Unknown anywhere. Not equivalent to [the Chinese] Sinian, which is known, and is considered a system that lies between the Cambrian and Precambrian by many, but not all, Russian authors."[28]

A contender for the name of the immediately preceding Precambrian period and system is the Sinian. The Sinian system was proposed in 1877 by Ferdinand von Richthofen. Later, the Sinian was proposed as both a system and a period of the late Precambrian by Grabau,[29] who also assigned Precambrian sediments from North America and Europe to the Sinian system. Interestingly, however, Grabau originally regarded the Sinian system as Paleozoic.[30]

The problem with using the Sinian in such a way is that, as currently used, the Sinian system and period includes too much rock and represents too much geological time, respectively, to properly serve as a geologic period and system. For instance, Michael E. Brookfield has the Sinian spanning from 800 to 530 million years ago.[31] This is 270 million years and is nearly quadruple the length of the longest existing geological period. Under Russian usage the Sinian has an even broader range, spanning from 1.2 billion to 570 million years ago.[32]

I propose a solution to this difficulty, one that will bring the Sinian into line with the approximate lengths of the other geological periods. I think that it is important to do so because our knowledge of the Proterozoic has advanced to the stage where it is pointless for us to claim ignorance about the Precambrian as an excuse for lumping vast amounts of geological time into formally named intervals. Instead, let us make the Lipalian period and system[33] represent the last part of the Precambrian (600 to 541 million years ago) and make the Sinian period and system the interval immediately before the Lipalian.

The total duration of the Sinian should be less than 100 million years, to keep it in accord with the lengths of other periods and systems. This will necessitate truncating the beginning and the end of the Sinian as currently used, but will have the important benefit of rendering the Sinian a globally used and useful unit of geological time. I will defer to my Chinese colleagues for selection of an appropriate type section and area for the Sinian system.

As for the Lipalian system, however, a North American stratotype is required. Walcott defined no stratotype for the Lipalian, so I propose that the Sonoran section of the Cerro Rajón, Mexico, be designated as the type section for the Lipalian.[34] This section has the advantages (once through a particular locked gate) of ease of access, excellent exposures of bedrock, abundant nearby correlative strata, and the fact that it has some of both the oldest Ediacaran fossils known and the oldest shelly fossils in North America.[35] The section is amenable to both radiometric and paleomagnetic dating and, if one knows where to look, abundantly fossiliferous.

Figure 10.1 shows the newly proposed subdivision of Precambrian time. A complementary system has been proposed by Hans Hofmann based on the Geon concept.[36] In the Geon concept Hofmann divides all geologic time into equal increments of 100 million years, assigning them numbers (Geon 1, Geon 2, etc.). He has even proposed a geologic map using Geon units, and perhaps in the spirit of the new millennium, Hofmann recommends calling each interval of 1 billion years a Gigennium.[37]

Such scaling might be very suitable for a geologic map of Rodinia (figure 8.4). The Geon concept is a useful supplementary system but must, in my opinion, be subsidiary to the traditional period and system framework because organisms and their fossil remains mark a unique signal in the geological record and radiometric age dates are subject to

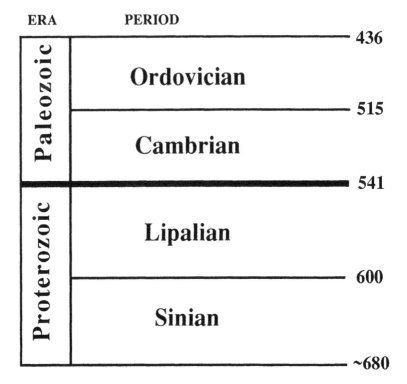

ERA **PERIOD**

	436
Paleozoic Ordovician	
	515
Cambrian	
	541
Proterozoic Lipalian	
	600
Sinian	
	~680

FIGURE 10.1: Newly proposed reorganization of geological periods in the vicinity of the Proterozoic-Cambrian boundary. All names for these periods have been proposed before, but never in this particular arrangement. The Paleozoic and Proterozoic eras extend beyond the amount of geologic time shown on this diagram.

recalculation and revision as techniques improve. Stratotype sequences based on bodies of rocks (as might be used to define system boundaries, for instance) do not change with continued research and must remain the fundamental criterion for geological time boundaries.

Clearly a number of pre-Sinian periods will be needed to fill out the Proterozoic, which began 2.5 billion years ago.

Notes

1. R. A. Fortey, "Review of M. A. S. McMenamin and D. L. S. McMenamin, 1990, *The Emergence of Animals: The Cambrian Breakthrough*, New York: Columbia University Press," *Historical Biology* 6 (1990):70–71.

2. See p. 217 in M. A. S. McMenamin and D. L. S. McMenamin, *Hypersea: Life on Land* (New York: Columbia University Press, 1994).

3. That is, not represented by sediment accumulation.

4. C. D. Walcott, "Cambrian Geology and Paleontology," *Smithsonian Miscellaneous Collections* 57 (1914):14. Walcott proposed the term *Lipalian* in 1910 but it was not published until 1914.

5. D. I. Axelrod, "Early Cambrian Marine Fauna," *Science* 128 (1958):7–9.

6. Page 146 in F. G. Snyder, "The Problem of the Lipalian Interval," *Journal of Geology* 55, no. 3 (1947):146–152.

7. Page 148.

8. Page 152.

9. W. K. Brooks had argued earlier (p. 457, "The Origin of the Oldest Fossils and the Discovery of the Bottom of the Ocean," *Journal of Geology* 2 [1894]:455–479) that "the evolution of animals likely to be preserved as fossils took place with comparative rapidity."

10. Page 173.

11. A. Seilacher, "Der Beginn des Kambriums als biologische Wende," *Neues Jahrbuch für Geologie und Paläontology, Abhandlungen* 108 (1956):155–180. Recently, however, the "Biogaeicum" has been extended into the Precambrian with the proposal of the biostratigraphic "*Dickinsonia costata* assemblage zone"; R. J. F. Jenkins, "The Problems and Potential of Using Animal Fossils and Trace Fossils in Terminal Proterozoic Biostratigraphy," *Precambrian Research* 73 (1995):51–69.

12. J. P. Grotzinger, S. A. Bowring, B. Z. Saylor, and A. J. Kaufman, "Biostratigraphic and Geochronologic Constraints on Early Animal Evolution," *Science* 270 (1995):598–604.

13. Similar thoughts were also expressed early on by Schindewolf; O. H. Schindewolf, *Der Zeitfactor in Geologie und Paläontologie* (Stuttgart: E. Schweizerbart, 1950).

14. Snyder, 1947, p. 150.

15. Fortey, 1990.

16. Which, as should be clear by now, I am paying close attention to in this book.

17. L. Lungerhausen, "On Certain Peculiar Features of the Ancient Series of the Western Slope of the South Urals," *Comptes Rendus (Doklady) de l'Académie des Sciences de l'URSS* 52, no. 2 (1946):159–162.

18. P. S. Pallas, "Observations sur la formation des montagnes et les changements arrivés au globe, particulièrement à l'égard de l'Empire Russe, *Acta Académie des Sciences de Imp. Petropolit. pro anno* (1777):21–64.

19. B. S. Sokolov, "On the Age of the Oldest Sedimentary Cover of the Russian Platform," *Izvestiya Acad. Nauk SSSR, Geol. Ser.* 5 (1952):12–20.

20. Page 28 in L. J. Salop, *Precambrian of the Northern Hemisphere and General Features of Early Geological Evolution* (Amsterdam: Elsevier, 1977).

21. Page 791 in P. Cloud and M. F. Glaessner, "The Ediacarian Period and System: Metazoa Inherit the Earth," *Science* 218 (1982):783–792.

22. W. C. Brögger, "Norges geologi," in *Norge i 19de Aarhundret, bd. 1, Kristiana, Central torylekeriet* (1900):1–32.

23. N. Menchikoff, "Quelques traits de l'histoire géologique du Sahara occidental," *Ann. Hébert et Haug.* 7 (1949):303–325.

24. R. B. Neuman and A. R. Palmer, "Critique of Eocambrian and Infracambrian," in J. Rodgers, ed., *El Sistema Cámbrico, su Paleogeografía y el Problema de su Base. XX Congreso Geológico Internacional, XX Sesión, México, Primer Tomo*, pp. 427–435 (Mexico City: Comisión Internacional de Estratigrafía and Unión Paleontológica Internacional, 1956).

25. H. Termier and G. Termier, "Ediakarskaya fauna i evolyutsiya zhivotnogo mira," *Paleontologicheskii Zhurnal* 3 (1976):22–29; translated into English as H. Termier and G. Termier, "The Ediacarian Fauna and Animal Evolution," *Paleontological Journal* 3 (1976):264–270.

26. G. O. Smith, *Charles Doolittle Walcott* (Washington, D.C.: Smithsonian Institution, 1927; privately reprinted from *American Journal of Science*, v. 14); E. L. Yochelson, "Discovery, Collection and Description of the Middle Cambrian Burgess Shale Biota by Charles Doolittle Walcott," *Proceedings of the American Philosophical Society* 140, no. 4 (1996):469–545.

27. Yu. R. Bekker, "Novoe mestonakhozhdenie fauny ediakarskogo tipa na urlae [A New Locality of Ediacaran-Type Fauna in the Urals]," *Doklady Akademii Nauk SSSR* 254, no. 2 (1980):480–482; Yu. R. Bekker and N. V. Kishka, "Otkritie ediakarskoi biota na yushnom urale [Description of the Ediacaran Biota in the Southern Urals]," in T. N. Bogdanova and L. I. Khozatsky, eds., *Teoreticheskie i prikladnye aspekty sovremennoi paleontologii [Theoretical and applied aspects of modern paleontology]*, pp. 109–120 (Leningrad: Leningradskoe Otdelenie "Nauka," 1989); Yu. R. Bekker, "Novyi predstavitel' drevneishei fauny urala [New Representatives of the Ancient Fauna of the Urals]," *Doklady Akademii Nauk SSSR* 310, no. 4 (1990):969–974.

28. Page 256 of the American Geological Institute, *Dictionary of Geological Terms*, rev. ed. (Garden City, N.Y.: Anchor Press/Doubleday, 1976).

29. A. W. Grabau, "The Sinian System," *Bulletin of the Geological Society of China* 1 (1922):44–88.

30. H. J. Hofmann, "Major Divisions of Earth History: Discussion," *Terra Cognita* 5 (1985):359–361.

31. M. E. Brookfield, "Problems in Applying Preservation, Facies and Sequence Models to Sinian (Neoproterozoic) Glacial Sequences in Australia and Asia," *Precambrian Research* 70 (1994):113–143.

32. See p. 5 in C. I Kuznetsov, M. V. Ivanov, and N. N. Lyalikova, *Vvedenie v geologicheskuyu mikrobiologiyu [Introduction to Geological Microbiology]* (Moscow: Izdatelst'vo Akademii Nauk SSSR, 1962).

33. The Lipalian system is a time-rock unit used to refer to all of the rocks formed during the Lipalian period.

34. M. A. S. McMenamin, S. M. Awramik, and J. H. Stewart, "Precambrian-Cambrian Transition Problem in Western North America: Part II. Early Cambrian Skeletonized Fauna and Associated Fossils from Sonora, Mexico," *Geology* 11 (1983):227–230; P. W. Signor, M. A. S. McMenamin, D. A. Gevirtzman, and J. F. Mount, "Two New Pre-Trilobite Faunas from Western North America," *Nature* 303 (1983):415–418; J. H. Stewart, M. A. S. McMenamin, and J. M. Morales, "Upper Proterozoic and Cambrian Rocks in the Caborca Region, Sonora, Mexico," *Physical*

Stratigraphy, Biostratigraphy, Paleocurrent Studies and Regional Relations, U.S. Geological Survey Professional Paper 1309 (Washington, D.C.: U.S. Government Printing Office, 1984).

35. M. A. S. McMenamin, "Basal Cambrian Small Shelly Fossils from the La Ciénega Formation, Northwestern Sonora, Mexico," *Journal of Paleontology* 59 (1985):1414–1425.

36. H. J. Hofmann, "Precambrian Time Units: Geon or Geologic Unit?" *Geology* 19 (1991):958–959; H. J. Hofmann, "New Precambrian Time Scale: Comments," *Episodes* 15 (1992):122–123.

37. H. J. Hofmann, "Major Divisions of Earth History: Discussion," *Terra Cognita* 5 (1985):359–361.

11 · A Family Tree

Considering their importance for our understanding of early animal evolution, the unresolved systematic questions will continue to dominate discussion of the Ediacaran fossils.

—Mark A. S. McMenamin[1]

If you are able to state a problem, it can be solved.

—Edwin H. Land, American inventor (1909–1991)

Ever since Seilacher's dramatic reinterpretation of the Ediacaran fossils, paleontologists have been embroiled in a heated controversy over the taxonomic affinities of the Ediacarans. Simply stated, what are these creatures? Are they weird animals, or do they represent something completely different, not animals at all? And when (or did?) they ever go extinct? Are we even asking the right questions here?

Foremost among the enigmas of paleontology is the taxonomic position of the Ediacaran soft-bodied fossils. What an intractable problem this case has been! A new hominid fossil or a new dinosaur almost always slides neatly into a fairly well structured conceptual framework. For hundreds of years, given a single fossil claw or tooth, paleontologists have been able to assume a significant amount of knowledge concerning the anatomy and habits of the deceased. But with a fossil "unknown," all bets are off for making such assumptions. How to proceed, then, with identification?

First, the fossils themselves give clues. These clues can be accessed by asking the right questions. How were the organisms preserved? How were the sediments that bear them deposited? In what type of environment were the creatures living?[2] Second, the shape and symmetry of the fossils provide further clues. Was the creature radially or bilaterally symmetric? Was its top similar to its underside? Does it have a thick body or is it flat (and was it flattened by compaction of enclosing sediment)?

A major problem with this type of study involves the repeatability, through the course of evolutionary change, of certain basic geometric patterns. For example, completely unrelated organisms can evolve similar shapes, and the shapes of the unrelated organisms can be so similar that it can sometimes be difficult to distinguish the different organisms even when they are alive and side by side. The difficulties are compounded when all one has for comparison are fossil remains of simple morphology. Branching patterns, discs with concentric rings, leaflike shapes, and bilateral, radial, threefold, fourfold, and other types of symmetries have appeared and disappeared again and again throughout the history of life.

Clifford E. Lundberg wrote in the late 1980s that "extensive serial homology is obvious in many early Paleozoic fossils and in the prior Ediacaran fauna as well; this suggests that segmentation formed rapidly."[3] Homeobox (Hox) or homeotic genes are responsible for limb patterning and other features genetically expressed along the axis of animals, including the parts of the brain. Rudolf A. Raff fluently describes how these genes play the same role in determining body forms such as the common development of the central nervous system of both arthropods and mammals.[4] Runnegar makes an interesting inference regarding homeobox genes in forms such as *Dickinsonia* and *Spriggina*, arguing that the presence of homeotic genes "in *Dickinsonia* and *Spriggina* may be inferred from their strict bilateral symmetry."[5]

As a general rule, paleontologists do not have access to the genetic makeup of the organisms they study, so in most cases they have little or no opportunity to compare the gene sequences of ancient organisms with their living relatives. The task of the paleontologist is thus to find morphologic characters or traits useful as the key to unlocking family histories. Such crucial traits are not uncommon, but great care must be exercised in their use and identification to ensure that the trait of interest can indeed be validly used as a family character and is not a structure developed independently by the putative ancestor-descendant pair.

Finding such traits involves the talent of the artist[6] as well as the talent of the scientist, especially when dealing with enigmatic fossils. Several layers of imaginative thinking must be used to tackle the most difficult fossil affinity problems, and in my opinion such problems are among the most, if not *the* most, difficult and demanding of all research problems in the sciences. The entire process is fraught with possibilities for error, yet these problems are squarely within the realm of science. These problems can be solved to the satisfaction of nearly every practicing paleontologist.

With that as introduction, you may have guessed that I am about to propose a solution to the problem of the affinities of some of the Ediacaran organisms. Indeed I am. The solution proposed here is highly testable and will stimulate further inquiry.

Allow me to begin with what I believe is the wrong answer. Ediacaran forms could be likened to a group of small marine animals called mesozoans. There are two small phyla of mesozoans: the orthonectids and the dicyemids. Orthonectids are parasites on a variety of invertebrate hosts (turbellarian flatworms, echinoderms, polychaete annelids, sea squirts, and gastropods), taking over the host's tissues with an internal, ameboid syncytium[7] filled with "seeds" of new parasites.

Dicyemids (figure 11.1) are also parasitic, attaching themselves only to the excretory organs of cephalopod mollusks. What looks like the animal's gut is actually an axial cell. This axial cell, which may occur in rows of three, is surrounded by ciliated jacket cells. There is a specialized organ, the calotte, at one end of the animal used for attaching to host tissue. As may have been the case for many Ediacaran body fossils, dicyemids have no mouth, gut, eyes, or nerve tissue. The jacket cells, which grow only by cell enlargement, could be likened to the pneu tubes in Ediacaran forms. The arrangement of jacket cells looks like a zigzag medial suture.

Dicyemids have been identified as models of what the first primitive animals might have been like. Some scientists have gone so far as to say that they are descendants of the "ancestral animal."

However, the interpretation of dicyemids as ancestral animals or Ediacaran survivors is probably wrong. Genetic sequencing of 18S rDNA

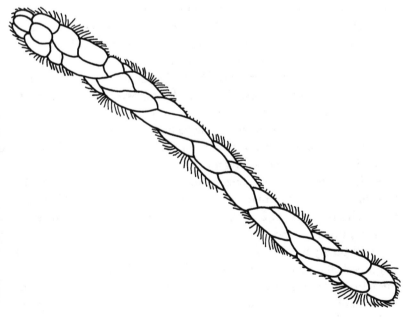

FIGURE 11.1: This sketch portrays a mesozoan animal, a member of a group called the dicyemids. Dicyemids are exclusively parasitic on the excretory organs of cephalopod (squid, octopus, nautilus) mollusks. Length approximately 0.7 mm.

in *Dicyema orientale* and *Dicyema acuticephalum*, parasitic species isolated from cephalopod urine, favor the hypothesis that dicyemids are not "an early divergent metazoan group, but rather a group degenerated from a triploblastic [that is, 'advanced' animal] ancestor."[8]

First, unless the Ediacaran forms are themselves independently degenerate from a triploblastic ancestor (an interesting possibility), Pflug and Seilacher could be right that Ediacaran forms may not be multicellular in the conventional sense. Any paleontologist who understands animals realizes that all animals must be multicellular because, as per the definition of animal, they must be able to form a multicellular blastula.[9] Could the Ediacarans have not been multicellular at all, but instead been giant unicells? Bruce Runnegar mentions the possibility that vendobionts were derived "from an unknown group of unicellular protists or aggregative amoebae."[10] He further noted that they could have belonged to any of a large number of protist phyla.

Rudolf Raff has command of the issues involved in trying to decide to whom the Ediacarans are most closely related. He describes the four main competing hypotheses proposed to explain the relationship between the Ediacaran biota and the animal family tree.[11] In the ancestral metazoans hypothesis, Ediacaran forms are seen as directly ancestral to all later animals. In the early diploblasts[12] hypothesis, Ediacaran creatures are also directly ancestral to living forms, but only to other diploblastic animals such as jellyfish and corals. In the garden of Ediacara hypothesis, a "now-extinct offshoot of pleated sheet animals probably living in symbiotic association with photosynthetic algae occupied the Earth before the rise of the hungry metazoans of the Cambrian radiation." And finally, in Seilacher's vendozoan hypothesis, Ediacaran creatures were not metazoans at all but some entirely different and now extinct type of organism.

Adding to the confusion, there is no scientific consensus on when the first animals evolved; estimates range from 1.2 billion to 600 million years ago.[13] This is a span of 600 million years, the difference in time between the first Ediacarans and today.

On February 4, 1997, I received a telephone call from my editor, Ed Lugenbeel, asking me whether I could deliver to him the manuscript for this book in 2 weeks. Wanting to meet this deadline, I scanned the status of the book project to see what remained to be done. Reviews were in, and most illustrations were ready or quickly obtainable. Just one problem. I still didn't have a satisfactory explanation for what the Ediacaran forms *were*.

I sat brooding in my office late that night (it was my birthday). I decided I needed to clear my head so I left the college and crossed the street to a coffee shop called The Thirsty Mind. Mount Holyoke students and local high school students were frequenting the place, chatting happily. I bought a cinnamon roll and a large espresso, kind of a weird combination, the sweet and the bitter. As I continued to brood by myself, the caffeine apparently took effect. It suddenly struck me that the key to the problem was the triradial nature of *Pteridinium*, as seen by viewing *Pteridinium* in cross-section (figure 4.3).

I dashed back to my office and pulled out my two specimens of *Pteridinium* from Namibia (color plate 8; figure 5.3). "Seilacher was on the right track," I thought to myself. "These things are weird, but they are not unicells. They are multicellular. But it is unusual multicellularity."

Pteridinium probably began life as a single cell. That cell divided to form two. Ordinarily in development the next step would be for both of these cells to divide, forming four. But suppose only one of them divided. This would give three cells. These three cells must, in some fundamental way, be responsible for the trifold shape of *Pteridinium* in cross-section.

I looked at the illustrations of Ediacaran fossils in Raff's book.[14] It suddenly became crystal clear to me how one is to consider the biology of the Ediacaran fauna.

The shape of Ediacaran organisms is controlled by the early cell divisions. Each cell of this early stage gives rise to descendant cells that might be called cell families or cell lineages.[15] However, unlike the cell lineages of other animals, these cell families tend to be quite independent and separate, and when they reproduce, they form cognate cell families. Thus the frond of *Pteridinium* elongates three cells at a time, and each of the three develops an individual tube or quilt of the right bathtub, the left bathtub, and the chaperone wall. Each quilt is multicellular, and is fused to adjacent quilts, but normally there is no mixing between cells of separate families.

This process explains the morphology of what I will call the one-cell, two-cell, three-cell, four-cell, and five-cell Ediacaran specimens. These can be further categorized as to whether the founder cell families give rise to offspring (cognate families of cells). In other words, if there are one-member cell families, then the total number of cell families in the mature Ediacaran organism equals the original number of cell families. If the cell families reproduce, however, then the total number of cell families increases with time as successive iterations of cell families are grown. These Ediacaran forms are the ones with long axes because the iterating cell families tend to stack up.

Table 11.1 shows how the Ediacaran taxa can be organized in this fashion. With a single cell family that does not reproduce additional cell families, the form develops into a simple, discoid *Cyclomedusa* (figure 2.1) or a *Beltanelliformis* (figure 7.2). With two cell families, something resembling the bilobe ?*Cyclomedusa* (figure 11.2) develops. This specimen bears an interesting similarity to the flattened, paired fleshy leaves of the desert succulent *Lithops* (color plate 4).

An even better example of two cell families is *Gehlingia dibrachida* (figure 2.11). With three cells and no iteration of cell families, *Tribrachidium* (figure 1.1) develops. Four cells give the quadripartite *Conomedusites* (figure 11.3). Five cell families yield the pentamerally symmetric *Arkarua* (figure 2.27).

When cell families iterate, elongate forms develop. When one cell family iterates, a *Charnia* or a *Charniodiscus* develops. The bulbous base of *Charniodiscus* is the "overgrown" initial cell family. Each alternating quilt of the frond represents a new cell family, formed at the tip of the frond by the immediately previously born cell family. Cell families enlarge with time. When two cell families iterate, a *Dickinsonia* is formed. Iterate three cell families, and *Pteridinium* is developed. This explains why *Pteridinium* is triradiate in cross-section; it is an expression of the original three cell families. The three families iterate over the life of the *Pteridinium*, extending its trifold body architecture.

With four iterating cell families, *Spriggina* (figure 2.19) is formed. The lower two cell families in *Spriggina* are programmed to elongate as in frond fossils such as *Charniodiscus*, whereas the upper two cell families remain subspherical. *Spriggina* appears to represent an exception to strict

TABLE 11.1

ORIGINAL NUMBER OF FOUNDING CELLS	CELL FAMILIES WITH FOUNDING MEMBERS ONLY	METACELLULAR CELL FAMILIES, UNIPOLAR	METACELLULAR CELL FAMILIES, BIPOLAR
One	*Cyclomedusa*	*Charniodiscus*	Vendofusa
Two	*Gehlingia*	*Dickinsonia*	*Windermeria*
Three	*Tribrachidium*	*Inkrylovia*	*Pteridinium*
Four	*Conomedusites*	*Spriggina*	?
Five	*Arkarua*	*Rangea* (?)	?
Six	?	*Swartpuntia* (?)	?

FIGURE 11.2: A possible two-cell-family Ediacaran, ?*Cyclomedusa*, from the Ediacara Member of the Rawnsley Quartzite, Ediacara Hills, Flinders Ranges, South Australia. The question mark in the name refers to the uncertainty entertained by paleontologists as to whether this specimen should actually be placed in the genus *Cyclomedusa*. Specimen is 6.6 cm in diameter.

segregation of cell families. At the "head" end of *Spriggina*, the founder cell families (and perhaps the first several iterations) have fused (comparable to the process known as tagmosis in arthropods) to form a shield-shaped structure. A similar type of fusion is seen in *Marywadea*,[16] a form resembling *Marywadea*[17] (two cell families, iteration), and other Ediacarans. The function of these fused cell families is not known, but it is conceivable that they served as some type of attachment device or cluster of sense organs that communicated with other cell family units along the axis of the body (where partial fusion of cell families may also have occurred).

Ediacaran fossils can now be organized, for the first time, into a cladogram, or family tree. The main division in the Ediacaran family tree is between forms with a fixed number of cell families and forms with iterating cell families. Along each of these trunks of the Ediacaran family tree, genera are arrayed depending on whether they display one cell fam-

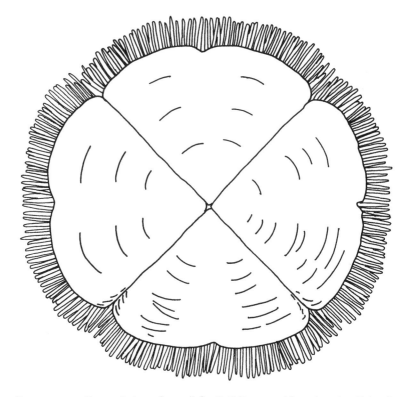

FIGURE 11.3: *Conomedusites*, a four-cell-family Ediacaran without iteration. From the Ediacara Member of the Rawnsley Quartzite, Ediacara Hills, Flinders Ranges, South Australia. Specimen is 7.4 cm in diameter.

ily, two cell families, three cell families, and so on. Such a classification would be imperfect because, for instance, Vendofusa (figure 7.1) has two iterating cell families ("bipolar" in the terminology of A. Seilacher), but they are on opposite ends of the animal, thus giving its characteristic spindle shape. Therefore, another distinction must be added to the classification: the distinction between whether the multicellular cell families are unipolar or bipolar (figure 11.4).

The cladogram shown in figure 11.5 differs somewhat from an ordinary cladogram in that the relationships between cell families may not be indicative of ancestor-descendant relationships but between shared geometries of cell family iteration. Changes in body characteristics are referred to by evolutionists as different grades of evolution; thus figure 11.5 might be more appropriately called a "gradogram." In fact, it may be a reasonable approximation of Ediacaran phylogeny; for instance, *Arkarua* and *Tribrachidium* are tribrachidiids on the same main branch

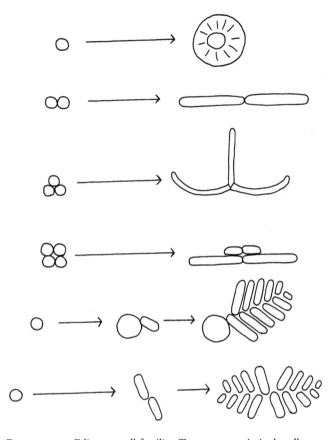

FIGURE 11.4: Ediacaran cell families. Top sequence: A single-cell propagule of a one-cell-family Ediacaran with no iteration develops into a medusoid such as *Cyclomedusa*. Second sequence: A two-cell propagule of a two-cell-family Ediacaran with iteration develops into a flattened form such as *Dickinsonia*, shown here in cross-section. Third sequence: A three-cell propagule of a three-cell-family Ediacaran with iteration develops into a trifold form such as *Pteridinium*, shown here in cross-section. Were a trifold form to develop without iteration, an organism such as *Albumares* or *Anfesta* (figure 2.10) would develop; the three large structures in the center of these forms represent the three founder cells (they are even somewhat elongate, as in *Dickinsonia* and *Pteridinium*). The center point of *Albumares* or *Anfesta* thus becomes equivalent to the zigzag medial suture in *Pteridinium*. Fourth sequence: A four-cell propagule of a four-cell-family Ediacaran with iteration develops into a flattened, bilaterally symmetric form. It is hypothesized here that such forms were able to cephalize by fusion of cell families into an anterior sense organ cluster. Fifth sequence: A single-cell propagule of a one-cell-family Ediacaran with iteration develops into a unipolar frond such as *Charniodiscus*. The founding cell family enlarges to form the base of the frond (recall again the enlarged three founding cell families of *Albumares* or *Anfesta*). Sixth sequence: A single-cell propagule of a one-cell-family Ediacaran with iteration develops into a bipolar frond (such as Vendofusa) by iterating new cell families in both directions. See also Table 11.1.

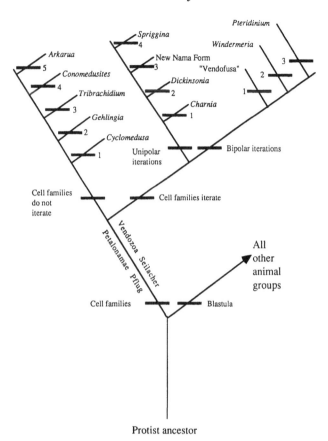

FIGURE 11.5: A family tree for Ediacarans, organized by number of cell families, whether the cell families iterate, and whether iterations are unipolar or bipolar. Ediacarans are related to animals (both share a common ancestry from unicellular protists) but do not go through the animalian blastula stage during development of the embryo.

The lowest Y-branch in this diagram can be called the Pflug/Seilacher dichotomy. Compare this with Banks dichotomy in the family tree of land plants; see M. A. S. McMenamin and D. L. S. McMenamin, *Hypersea: Life on Land* (New York: Columbia University Press, 1994).

and probably were closely related. At a minimum, this gradogram/clado-gram may be used as a starting hypothesis against which other schemes of Ediacaran phylogeny may be tested.

What I conclude from this new analysis is that the creatures of the phylum Petalonamae Pflug or phylum Vendozoa Seilacher, the Ediacaran organisms, were closely related to "normal" animals (the metazoa), shared a common ancestor with them from a protist[18] ancestor, but developed a highly unusual approach to multicellularity that seems to have features in common with modern xenophyophores (such as sand within the body and periodic growth increments), those strange, large marine protists described earlier in this book.[19]

But I do not think that the Ediacaran fossils were xenophyophores or any other type of protist. Ediacaran body form is much more suggestive of animals (although large Protista may have been present in the same environments); furthermore, these body fossils appear at the same time as the earliest undoubted animal fossils.[20] There is likely to be a close genetic relationship between the animals and the Ediacarans. To my mind, in this case shared time of origin implies shared descent.

The Ediacaran body was composed of one or more cell families. Each separate cell family was packaged in some sort of resilient integument, which may help explain why these "soft-bodied" organisms preserved in coarse sediment. Each cell family, delineated by the cuticle, may have thrived in a semiautonomous manner, although evidence for cell family fusion in frond forms and for coordinated cell family contraction in *Dickinsonia* suggests at least a limited degree of communication between cell families within the same organism.

Humberto R. Maturana and Francisco J. Varela call this type of organization *metacellularity*. They define metacellularity as any organism "in whose structure we can distinguish cell aggregates in close coupling."[21] For the Ediacarans, the cell aggregates in the sense of Maturana and Varela are here called the *cell families*.

Thus Pflug was approaching the right track when he suggested that the Petalonamae were colonial.[22] Ediacaran forms do not show true coloniality, but rather coordination between cell families. Included among the constituent cells of the cell family were presumably chemo-symbionts and photosymbionts, and in some species, sand taken from the surrounding environment. Alternatively, symbionts were passed along with each successive division of a cell family. Individual cell families could be modified for photosymbiosis. Note the distal[23] flattening of the lower (elongate) pairs of cell families in *Spriggina*.

Reproduction in Ediacaran forms was simple; cells merely had to split off from the growing end or ends. In an iterated form, every, say, 10 reproductions of cell families would alternate with release of tiny cell families to found new individuals. These cells may even have come off in clusters, two for *Dickinsonia*, three for *Pteridinium*, and so on.

I have cracked the code of the Ediacaran puzzle.[24] Next time you need to solve a problem, order a cinnamon roll and an espresso, preferably at The Thirsty Mind.

Notes

1. M. A. S. McMenamin, "The Dawn of Animal Life," *Episodes* 11 (1988):229–230.

2. There is a wonderful German word, *Umwelt*, that refers to the environment as perceived by an organism. See F. DeWall, "Are We in Anthropodenial?" *Discover* 18 (1997):50–53.

3. Page 32 in C. E. Lundberg, *Segmentation: A New Evolutionary Theory* (San Francisco: Glonora Press, 1988). Lundberg defined serial homology (p. 58) as the "condition in which homologous structures (structures which are similar due to common ancestry) are arranged in linear order, forming a chain of segments. If all segments are identical [as in certain members of the Ediacaran biota], the serial homology is ideal."

4. R. A. Raff, *The Shape of Life: Genes, Development, and the Evolution of Animal Form* (Chicago: University of Chicago Press, 1996).

5. See p. 313 in B. Runnegar, "Vendobionta or Metazoa? Developments in Understanding the Ediacara 'Fauna,' " *Neues Jahrbuch für Geologie und Paläontologie, Abhandlungen* 195, nos. 1–3 (1995):303–318.

6. See p. 10 of E. M. Moores, "Geology and Culture: A Call for Action," *GSA Today* 7 (1997):7–11.

7. A syncytium is an acellular body mass filled with unattached cell nuclei.

8. See p. 81 of T. Katayama, H. Wada, H. Furuya, N. Satoh, and M. Yamamoto, "Phylogenetic Position of the Dicyemid Mesozoa Inferred from 18S rDNA Sequences," *Biological Bulletin* 189 (1995):81–90.

9. An animal is defined here as a multicellular creature with a blastula-forming embryo and anisogametous (male/female sex cell) reproduction.

10. See Runnegar, 1995, p. 310.

11. See figure 3.2, p. 72 in R. A. Raff, *The Shape of Life: Genes, Development, and the Evolution of Animal Form* (Chicago: University of Chicago Press, 1996).

12. A diploblast animal has two cellular body wall layers. Examples include jellyfish and sea anemones.

13. J. W. Valentine, D. H. Erwin, and D. Jablonski, "Developmental Evolution of Metazoan Bodyplans: The Fossil Evidence," *Developmental Biology* 173 (1996):373–381; G. J. Vermeij, "Animal Origins," *Science* 274 (1996):527–528; G. A. Wray, J. S. Levinton, and L. H. Shapiro, "Molecular Evidence for Deep Precambrian Divergences Among Metazoa Phyla," *Science* 274 (1996):568–573.

14. His figure 3.1.

15. L. W. Buss, *The Evolution of Individuality* (Princeton, N.J.: Princeton University Press, 1987).

16. M. F. Glaessner, "A New Genus of Late Precambrian Polychaete Worms from South Australia," *Transactions of the Royal Society of South Australia* 100 (1976):169–170.

17. Plate 4, figure 1 of J. G. Gehling, "The Case for Ediacaran Fossil Roots to the Metazoan Tree," *Geological Society of India Memoir* 20 (1990):181–224.

18. Choanoflagellate.

19. Indeed, the xenophyophore *Stannophyllum zonarium* looks much like an Ediacaran medusoid.

20. M. A. S. McMenamin, "Ediacaran Biota from Sonora, Mexico," *Proceedings of the National Academy of Sciences (USA)* 93 (1996):4990–4993.

21. See p. 87 in H. R. Maturana and F. J. Varela, *The Tree of Knowledge: The Biological Roots of Human Understanding* (Boston: Shambhala, 1987).

22. H. D. Pflug, "Zur Fauna der Nama-Schichten in Südwest-Afrika. III. Erniettomorpha, Bau und Systematik," *Palaeontographica* 139A (1972):134–170.

23. Meaning here "outward," "away from the central axis."

24. I first publicly announced the metacellularity/cell families hypothesis at the annual Geological Society of America meeting in Salt Lake City, Utah, October 1997 (M. A. S. McMenamin, "Metacellularity, Cognate Cell Families, and the Ediacaran Biota," *Geological Society of America Abstracts with Program* 29 [1997]:A-30). The presentation generated considerable interest from colleagues. As I wrote to Dave Evans (on October 29, 1997) in reply to his questions regarding metacellularity and his correction of the provenance of one of the slides which I had used as an architectural analogy to the idea of cell families/metacellularity:

> You're half right, my slide projected on the left side was indeed from the Casa Battlo. The one on the right was from La Sagrada Familia, showing the back side of one of the tower tops (in a style very much like Battlo). Thanks for the correction. (See G. R. Collins, *Gaudi*, 12th ed. [Barcelona, Spain: Editorial Escudo de Oro, 1990]).
>
> My main argument is this: the somatic compartmentalization of the Ediacarans is so profound that it has to represent something very basic about their early ontogeny/embryology. Instead of going through a blastula stage (with lots of coordinated cell movements), we are seeing something like "selfish [cell families]." In other words, the business end of the nipple is an overgrown [cell family] that finally, and perhaps reluctantly, allows differentiation to occur to form the radial stuff on its perimeter. . . . *Tribrachidium* does the same, except this time there are three selfish [cell families]. For an *Inaria* critter (looks much like the five-armed cross motif at Battlo) we are dealing with five to eight garlic clove shaped selfish [cell families]. Perhaps this is enough "body" now to not require the radial [elaboration]. And I bet the propagules formed right at the tip, each with its complement of founding cell families, just as [Spanish architect Antonio] Gaudi has it at Battlo. This metacellularity idea is really working for me as a means of understanding the biology of these things, and I have several ideas on how to test it.

The idea that Ediacarans can have both a regularized, iterated cell family part *and* a radial/tubular outgrowth from one or more of the cell families, perhaps as a means of increasing surface area, would be confirmed by finding, say, a specimen of a dickinsoniid with branching radial structures on its margin. In fact, such a specimen exists and was described in 1973 (G. J. B. Germs, "Possible Sprigginid Worm and a New Trace Fossil from the Nama Group, South West Africa," *Geology* 1 [1973]:69–70). This specimen has numerous iterated cell families *plus* tribrachidiid-like radial structures around its margin.

Another recent discovery shows this combination of the main cell-family part of the Ediacaran body with a radial part formed of branching canals. Newly described *Ventogyrus chistyakovi* (A. Yu. Ivantsov and D. V. Grazhdankin, "A New Representative of the Petalonamae from the Upper Vendian of the Arkhangelsk region," *Paleontological Journal* 31 [1997]:1–16) is a short, boat-shaped, three-dimensional form similar in some respects to *Pteridinium*. New cell families were added at the "bow" of the boat; a triangular sternal chamber called the *camera puppe* occurs at the "stern." This triangular structure, which is homologous both to the funnel-shaped "prostomium" in Germ's fossil mentioned in the previous paragraph and to the "head" in *Marywadea*, is the source in *Ventogyrus* of branching canals that form a covering sheet over the sternal part of the underside of *Ventogyrus*. Thus the fossil is like a *Pteridinium* enveloped in the "head" region of a *Marywadea*. The function of these canals is unknown, but they might have some sort of sensory or trophic function. I am reminded of the division of chytrids into a basal cell (cf. founding cell family) and fine rhizoids (cf. radial part of an Ediacaran body). The fact that all the canals diverge from a main trunk canal in the camera puppe and that all canals hug close to the bottom surface of *Ventogyrus* (rather than spreading out into the sediment like fungal mycelia) may argue in favor of the sensory function hypothesis over the trophic function hypothesis for the canals. One might speculate that the array of canals, like the frondose antennae of moths, was intended for chemical communication (via sediment pore waters) with other ventogyrids.

I went on to explain in the lecture in Salt Lake City how the Ediacaran *Kimberella*, lately interpreted as Proterozoic mollusk (M. A. Fedonkin and B. M. Waggoner, "The Late Precambrian Fossil *Kimberella* Is a Mollusc-like Bilaterian Organism," *Nature* 388 ([1997]:868–871; R. Mestel, "*Kimberella*'s Slippers," *Earth* 6 [1997]:24–31), is no animal at all but rather a series of cell families, stacked on top of one another instead of forming a horizontal series of cell families spread out across a bedding plane surface. Cell families in *Kimberella* were generated or iterated from the "bottom" of the stack. Likewise, Seilacher's hypothesized molluscan "mat scratcher" and its putative radular markings represent, in fact, a cluster of Ediacarans in which the radial parts of the organism radiate outward from the initial cell family to form a fan pattern. (See p. 26 in A. Seilacher, *Fossil Art* [Alberta, Canada: Royal Tyrrell Museum of Paleontology, 1997]). The annulations of the central axis of *Swartpuntia* could easily accord with this model of vertically stacked arrangements of cell families.

12 · Awareness of Ediacara

And there were certain living things that were without perception, from which came others with perception, and they were called *Zophe shamin*, that is, Watchers of the Sky.

—Sanchuniathon, c. 600 B.C.[1]

Someone has to propose ideas at the boundaries of the plausible, in order to so annoy the experimentalists or observationalists that they'll be motivated to disprove the idea.

—Carl Sagan[2]

Cada cabeza es un mundo.[3]

—Anonymous

My first and only extended meeting with Steven Jay Gould occurred on Saint Patrick's Day, March 17, 1988, when I took my "Great Ideas in Geology" class to meet with him at Harvard. The meeting was not long (one hour) but was enjoyable and, at least for me, important. Gould, who had graciously agreed to meet with my class, asked whether I was having the students read original sources. As it turned out, I had an exceptional group of students that year in my seminar, and not only were they reading original sources, but Jennifer Convey (class of 1989) was reading Steno's *Prodromus* in the original Latin.[4] Steno had founded modern stratigraphy and was first to demonstrate the biological origins of fossils.

During our discussion I mentioned one of the essays in which Gould refers to human ecology as a contingent fact of history, and Gould's eyes lit up and he gave an involuntary jerk at the mention of the word *contingency*. I didn't know it at the time, but Gould was basing his current writing project on an attempt to underscore the importance of contingency in the evolutionary history of life.

The main focus of Gould's work, as exemplified by his *Scientific American* article[5] "The Evolution of Life on Earth" (the one with my *Pteridinium* photo), has been to curtail the thought that there is anything "progressive" about the evolutionary process.[6] In Gould's worldview, random (contingent) events control evolution and render the process unpredictable. If one could rewind the tape of life and play it over again, says Gould, nothing like human life (or even intelligent life) would have much chance of evolving a second time. Gould has defined intelligent life as a life form capable of understanding its own evolutionary history, a definition with which I would agree.

Gould's view on contingency is a minority one, however, and others have spoken out against this rendering of life's history. Simon Conway Morris, in his typically colorful language, argued that similar environmental conditions often cause unrelated organisms "to find the same biological solutions" because there are a limited number of ways that things can be done.[7] If the tape of life were to be rewound, the probability that any one of us would be here today is "infinitesimally small. But I'd say that the odds of an upright, two-legged, introspective organism are rather high."

Could such a scenario occur? To gain a fuller understanding of this question, I must first sketch out the background of evolutionary thought on the origins of animals. Yet again, we must turn to the scientific thought of German scientists. Otto Heinrich Schindewolf, in his 1950 book *Der Zeitfactor in Geologie und Paläontologie* (*The Time Factor in Geology and Paleontology*), spoke of how there were no forerunners of the Cambrian animals, but that they suddenly appeared as a result of *Grossmutation*, or rapid macroevolutionary change.[8] Seilacher in 1956 agreed with Schindewolf's assessment of the evolution of the Cambrian fauna (not surprising, because Seilacher was Schindewolf's student), but Seilacher went further, noting that the trace fossil makers of the Lipalian suddenly and simultaneously changed their behaviors and activities.[9] The Petrogaeicum gives way to the Biogaeicum.

Daniel I. Axelrod attacked the Schindewolf-Seilacher rendering of the Cambrian lower boundary, saying that their ideas were not "acceptable because they do not conform to our present understanding of the evolutionary process."[10] Here Axelrod was defending what is called the "modern synthesis" of neo-darwinian evolutionary thought.

Neo-darwinian thought not only is at odds with the abruptness of the Cambrian evolutionary event, but also "excludes symbiogenesis except as an oddity of limited interest, mainly to cell biologists and biochemists."[11] Now, thanks to our progress in understanding the Ediacarans, we can navigate past the Scylla of sclerotic neo-darwinian thought *and* the Charybdis of denial of progressive evolutionary change.[12]

A variety of exceptional types of organisms evolved during Lipalian times. Three major groups can be recognized, which I call the first round of advanced eukaryotes:

Metazoans (kingdom Animalia): the trace makers, perhaps Ediacaran giant tube worms, sponges.

Phylum Petalonamae (the Vendobionts, related to kingdom Animalia): *Ernietta, Pteridinium, Charnia, Rangea,* Vendofusa, *Dickinsonia, Spriggina, Marywadea, Arkarua, Windermeria, Tribrachidium, Albumares, Gehlingia, Parvancorina,* most specimens of *Cyclomedusa,* and most other Ediacaran body fossils.

Xenophyophores (kingdom Protista): dumplings, possibly some of the Ediacaran medusoids, particularly those showing pronounced growth banding.

It is not surprising that both animals and protists would evolve one of the simplest forms (medusoid) independently. It makes eminent good sense that animals, in their early stages of evolution, would be part of a plexus of evolving, related forms all of more or less equal rank. When brainy, motile animals appeared, the members of other kingdoms were driven into deeper water or slowly driven to extinction. Xenophyophores continue to live in great abundance today, but only in areas of very deep (500 m or more) marine waters. Their mode of feeding is still unknown but probably involves osmotrophy.

The curious flattened shapes of *Marywadea, Bomakellia, Spriggina, Mialsemia,* and "soft-bodied trilobite," and their shallow water habitats, probably indicate photosymbiotic lifestyles. But these genera also appear to have "heads." I interpret this to mean that Ediacarans were on the verge, by the end of the Lipalian period, of developing forms with cephalized bilateral symmetry: anterior sense organs and a brain formed by cell family fusion. *Spriggina* and *Marywadea* are essentially encephalized versions of *Dickinsonia*. This evolutionary development was *totally independent of metazoans.*

An iterative evolution[13] is thus implied for the brain-controlled body. With Ediacarans, it may have involved sensory systems totally unknown to us, which might have been of great use to a sessile creature. The "head" in *Spriggina* and the "eye ridge" in soft trilobite are organs of chemodetection (or other sensory function) and their associated nerve clusters.[14] "Soft trilobite" cannot be a true trilobite; its "eye ridge" is a single (not paired) *D*-shaped structure. "Soft trilobite" is thus a Cyclops and thus quite unlike any known trilobite. The amount of body devoted to these Ediacaran sensory functions is, ironically, greater than in ordinary animals. In his original description of the Ediacaran species *Marywadea ovata,* Martin F. Glaessner noted that its "head was conspicuous and relatively larger than in any known annelid."[15] This is particularly true when

Marywadea is compared with primitive animals such as the Cambrian lobopodian animals, which were virtually headless.[16] If you will permit me a pun, is it possible that, for the duration of the Lipalian period, Ediacarans were ahead of their time?

Why would any of the Ediacarans acquire a brain? Most of them apparently thrived without one. And because all of these forms were apparently immobile and blind, what need would they have for a centralized nervous system? As Humberto R. Maturana and Francisco J. Varela state, "from the standpoint of the nervous system's appearance and transformation, the possibility of movement is essential," and it is in "the establishing of motility that the nervous system becomes important."[17] But perhaps Maturana and Varela are being too conventionally "metazoocentric" in so tightly linking the development of brains to motility. The Ediacarans may have had different reasons for developing a central nervous system.

Such a sensory development could occur in organisms that feed by osmotrophy or photoautotrophy. This is why they look eerily familiar; they were "trying" another, albeit rudimentary, version of encephalization. Such things as timing of gamete or propagule release, regulation of temperature, anticipation of salinity changes, hardening the cuticle in preparation for a coming storm, triggering of chemical potentials on the surface of the organism to absorb dissolved food or admit light,[18] orientation with regard to light or sources of nutritious dissolved chemical sources,[19] and activation of internal pigments[20] to optimize conversion of various spectral components of sunlight into useful energy might be facilitated by a central nervous system, imparting a distinct survival advantage to Ediacarans of the peaceful but not entirely competition-free Garden of Ediacara. Convergently evolved eyes could also be part of the Ediacaran sensory mix. Recall the unusual lensless eye strips of the vent shrimp *Rimicaris*.[21] Ediacarans may represent an unusually advanced type of photoautotrophy. Not surprisingly, all of the cephalo-Ediacarans lived on the sediment surface, where they were in direct contact with the information streams of the water column.

Instead of organizing muscular motion, as in metazoans, Ediacaran brains may have coordinated chemical signals, both released into and detected within the watery milieu of seawater. Once again, the medium becomes the message. Some Ediacarans apparently developed sophisticated methods of interpreting aqueous geochemical conditions. One could think of it as a Vernadskian[22] chemocognition, the beginning of a chemonoösphere.

Can centralized sensory systems use chemocognition? Indeed they can. Metazoans developed such systems as they moved, long after the Cambrian, into the unfamiliar and hostile environment of land. Animal parasites on land, such as the schistosome flukes, live within the tissues of other animals. These flukes have developed sense organs that do not exist in their free-living relatives. And of course, canine olfactory abilities seem amazing by human standards.

Several questions must now be asked. Does chemodetection in land animals represent a re-evolution of capabilities first developed by Ediacarans? In what directions might the chemonoösphere have led, had not Ediacaran evolution been cut short by the animals of the Cambrian explosion?

My explanation for the cephalo-Ediacarans[23] is the only way to adequately explain the cephalized, animaloid shapes of some members of the fauna; the decidedly nonanimalian body architecture of most Ediacarans; the lack of evidence for mobility in Ediacarans; and the probably photoautotrophic, chemoautotrophic, or osmotrophic lifestyles of these organisms. At last we can understand the Ediacarans in relation to ordinary animals. The radiation of Ediacarans produced morphological analogues resembling some types of modern animals (for example, Ediacaran medusoids are equivalent to jellyfish; cephalo-Ediacarans are equivalent to familiar types of metazoans), but with a fundamentally different body architecture that developed within a unique ecological setting. The Ediacaran *umwelt* and the animal *umwelt* are profoundly different.[24] We can test my idea by continued careful study of Ediacarans. If some of the forms can be shown to have metazoan-style paired, lens-forming eyes, then my arguments must be rethought.

Metazoans supplanted the Ediacarans in the Cambrian, the sessile Ediacaran forms apparently becoming easy prey to the Cambrian predators. Is this because brains coordinating muscular contractions are inherently superior to organisms of a chemocognition-based biosphere? Or was it just as likely for the Ediacaran forms to gain ascendancy and gain control of the world through geochemistry, had they not lost out due to poor timing or bad luck?[25]

The emergence of animal predators may have led to overgrazing of the Garden of Ediacara but it had a tonic effect on the evolution of other animals, who quickly exploited the evolutionary potentials of newfound behaviors and skeletons, exactly the *Grossmutation* Schindewolf described. The *Grossmutation* was a direct result of near simultaneous challenge and opportunity. And as Snyder in 1947 and Dianna and I in *Emergence of*

Animals correctly stated, the emergence of skeletonized animals (and perhaps the evolutionary deployment of most animal phyla) took place in only a few million years. Seilacher's change in burrower activity is one of the shock waves felt as a result of this Cambrian explosion.[26]

Fedonkin analyzed the relationship between trace fossils of the Proterozoic and later and the behaviors exhibited by the animals that made them.[27] He noted that the increase in behavioral complexity was a discontinuous process and that elaboration of trace complexity was the result of two factors. The first was appearance of fundamentally new types of trackways (the first sinusoidal burrow, first spiral burrow, first branching burrow, and so on). The second was a combination of two or more of these fundamental types to form complex traces that could help the tracemaker tap into new food sources. For example, the combination of spiral trace and sinusoidal trace to form spiral-sinusoidal trace could lead to more complete (and hence efficient) surface deposit feeding of sediment (figure 12.1). Organisms with large enough brains or flexible enough behavior to engage in combinations of burrowing behaviors would potentially have considerable selective advantages over forms locked into a single burrowing pattern forever. Cephalo-Ediacarans may likewise have been exploiting and coordinating combinations of feeding strategies (light capture, chemosynthesis, diffusion feeding).

The feedback between eye and brain, vitally important in the Cambrian, manifested itself with even greater force (dare I say, came into better focus?) later on with the first appearance of binocular vision. In binocular vision paired images were united in the brain to provide more information about the environment than was available from the separate, nonoverlapping images. Greatly enhanced depth perception was an important result of this new development. The brain was required to become more adept in the processing of dimensional data, and a new type of intelligence was born.

This is why the entire fabric of life was altered with the Cambrian event. Light and access to light remained important, but for new reasons. The photosymbiotic Garden of Ediacara, with Ediacarans using their entire bodies as collectors for light and chemicals, was overturned by the Cambrian ecosystem, where animals concentrated their light sensitivity into paired, brain-tethered organs of vision. The old Ediacaran world gave way to a burst of evolutionary change—a burst that generated most of the world's animal phyla in an astonishingly brief moment of geological time.[28] All features of the living world, from rock-mimicking plants of Namibia (*Lithops erniana*) to human intelligence itself, bear the mark of

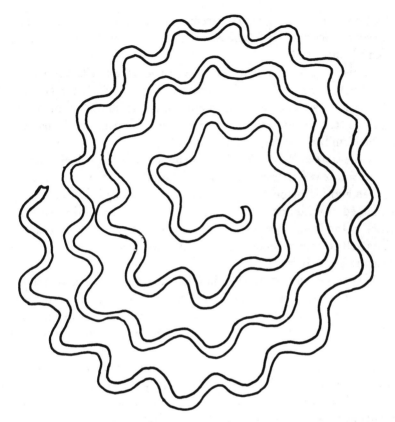

FIGURE 12.1: An ancient burrow or trace fossil formed by a combination of a sinusoidal burrowing behavior (imparting the wavy curve) and a spiral burrowing behavior (giving the trace its overall shape).

both the fall from the Garden of Ediacara and the rise of living things with brains. None bear this mark so conspicuously as the animals themselves.

Ediacarans thus are not too distantly related to animals[29]; they shared a common protist ancestor and (as is often the case in examples of iterative evolution) shared many of the same evolutionary potentials as animals. This then was the first radiation of "advanced" eukaryotes. Animals were not necessarily destined to win from the start, but somewhere in the Ediacaran-metazoan plexus was a group or groups with the potential to develop sentient life. Such a strategy is just too good a biological bet to pass up as soon as advanced, multicellular eukaryotes appear on the scene. And in the evolution of brains, survival advantages accrue to the cephalizers every evolutionary step of the way.

Let us return once more to the seemingly inextricable triad of German adventure, German wars of the twentieth century, and brilliant German scientific thought.

On the Air Namibia flight back to Germany, Herr Seilacher handed me a book, translated from the German, titled *The Sheltering Desert*.[30] The book is by Henno Martin, whom you will recall was an important figure in early Namibian geology. His picture still hangs in the hallway of the Namibian Geological Survey building.

At the start of World War II, two young German geologists working in South West Africa decided to make a dash for freedom. Henno Martin and Hermann Korn faced internment by Allied forces if they remained in Windhoek, so they elected instead to live a primitive existence in the Nama desert. They remained at large for two-and-a-half years, reluctantly surrendering to the authorities after Korn fell ill due to a vitamin deficiency.

After they turned themselves in (and Korn recovered), a Windhoek magistrate fined them for illegal possession of a radio, for an expired license for their truck,[31] and other charges, but perhaps because the judge felt sympathy for their plight, the fines were modest, and Martin and Korn were able to pay them with funds loaned by friends.

Martin and Korn were not Nazis; in fact, both men had fled Germany before 1938 because they abhorred what they felt was the madness of war. Both saw the war coming and sought to escape it. However, world events were soon to overtake them in their Nama refuge.

The men and their dog, Otto (so called because the name, like the dog itself, looked virtually the same from either end), mastered the challenge of desert wilderness living. Encounters with gemsbok, hyenas, zebra, pet fly-catching lizards, and the cultivation of desert radishes provided them with food, taught them in their times of scarcity why humans have a "fat tooth," and imparted to them an appreciation of, and a oneness with, the flow and ebb of life in the desert.

The two men had ample time for rambling discussion, and often the topic shifted to the meaning of evolutionary theory in a war-torn world. Martin narrates the key section beginning on p. 227:

> All these and similar questions to which we could find no satisfactory answers confirmed our feeling that no purely mechanistic interpretation, and in general no interpretation which assumed that a living being was only a complicated physio-chemical machine, was adequate and that applied not only to human beings, but to life

as a whole. So far so good, but where was the proof that there was a power superior to chemical and physical phenomena? We tackled the problem first from one angle and then from the other, but we made no progress. It was quite true that many of the almost incredible performances of animals and plants could be adequately explained by highly developed and perfected chemical and physical reactions operating entirely within material limits, but the fact remained that day after day our life on the edge of the desert demonstrated the fundamental difference between living and non-living matter.

One evening Hermann raised the subject once more: "Why does evolution only suggest that the problem of life can't be satisfactorily solved on a purely materialistic basis? Why doesn't it provide us with some proof? Perhaps there is no absolute proof," he continued, "because the accident of arbitrary mutation is embodied in evolution. Where chance is a condition of the experiment you can't demand a definite result. Chance is innate in the generational change. To avoid chance and thus the calculus of probability there would have to be evolution without either birth or death."

Suddenly the solution dawned on me.

"You said that if there were evolution without generational changes," I began. "But there is such a development where the individual man or animal is concerned. Every human being and every animal develops during the course of its life."

"True enough," admitted Hermann, "but can you prove that evolution isn't purely mechanically determined by the interplay of heredity and environment?"

"Yes, I think so. Take our lizards, for example. In just three days they learnt that a whistle meant food. The utmost their predecessors can have learned is that a buzzing noise indicated the approach of a fly. But the introduction of a connection between a whistle and the feeding reaction of the lizard required a readjustment and an extension of inherited characteristics. And such a readjustment and extension is what we call development, and in this case it was certainly no question of a mutation.

"And such readjustments happen constantly. In the Stone Age there was a man who was the first ever to trace the outline of an antelope in the sand. In doing so he was not only using one part of his body in a way it had never been used before, but he was deliberately creating something entirely new, to wit, a picture.

Before that no such thing as a picture had ever existed-not even as an idea. Thus life was not merely able to adapt itself to its surroundings, but it could create new things and give old things a new significance. And for this neither a generational change with accompanying mutations nor any natural selection in the struggle to survive was required.

"Now that certainly fits in with the conclusions we have already drawn regarding human evolution. We decided that it was precisely the protection of children from the struggle to survive which favored learning and new developments. But where any form of animal life is subjected too intensely to the struggle to survive we find narrow physical adaptations and rigid instinctive reactions which make further development impossible. Darwin was quite right when he said that the struggle for existence with the consequent survival of the fittest was responsible for adaptation in the animal kingdom. But as most adaptations hamper the capacity for further development their influence on life in general is negative, which is, after all, precisely what you would expect from a force operating primarily through destruction. Everything which is narrowly adapted to certain conditions are constantly changing [sic]. But where does that lead us? Above all it explodes all those wretched arguments which try to justify war and brutality on the ground that a ruthless struggle for survival furthers development. It furthers certain developments, but it drives them into a blind alley and leads ultimately to extinction.

"In other words, the feeling which revolted against such an idea was right, which isn't so very surprising because feeling is a judgment based on life as a whole whereas consciousness and understanding are newly acquired and not as yet highly developed faculties." . . .

This seemed to be an important realization which raised learning above all purely physical phenomena, for good and bad are not physical qualities even when they apply to physical things. And we could now link up this conclusion with some earlier ideas of ours when we decided that every feeling, even the simplest such as the perception of pain, involved a passing of judgment. Doesn't every feeling judge an experience in relation to a need? For example, the disappointment of a thirsty man when he finds that the water he wants to drink is salty. Or the feeling of relief a man experiences in the cool shade of a hillside out of the burning sun. Could one in fact go further and say that need was first made conscious by feeling?

More and more ideas developed. Was the capacity to pass judgment a spiritual attribute? If it were then a spiritual force was at work in all feelings, even the simplest. But was the expression "force" an appropriate one in this connection? Probably it was, because a feeling sent electrical impulses through the nervous system to set muscles into movement; in other words, it produced effects which could be produced only by forces in the strictly physical sense of the word. But the decisive thing was nevertheless that the forces were guided by judgments based on feeling. But feeling considered as a judgment does not necessarily run parallel with physical phenomena; it can run counter to them; it can approve or disapprove—in other words, it is superior to physical phenomena.

I had often asked myself whether life in all its multifarious forms was merely a strange accident in world evolution, and now I felt that I had obtained proof that it was nothing of the sort; life was raised above the phenomena of the inanimate world precisely by feeling. Here then was something truly remarkable: a force which determined values, something non-material, and guided physical phenomena through them. It was thus a force of a higher order, and by establishing values it broke through the rigid framework of purely physical phenomena.[32]

Martin's thought in these passages has close affinities to the *élan vital* camp of Bergsonian, vitalistic philosophy. The parallels with Teilhard de Chardin and Hans Driesch, other Europeans of their generation, are clear. Perhaps this Bergsonian sentiment is one born of saturation with and repulsion to the horrors of the world wars.

The criticisms that have been made of this progressive view of evolution are based primarily on rejection of biological teleology, direction toward a goal by biological change. It is impossible, however, to completely purge biology of teleology.

We will always have the ultimate *teleos* in biology, the fact that one bacterium will strive to become two, the fundamental aim of all living processes. Vladimir Vernadsky called this the "pressure of life," the observation that life will expand as far as it can even if it alters a planetary surface in the process.

So what of the Bergsonian noösphere, the thinking layer of the earth reified by LeRoy, Teilhard, and Vernadsky? The original concept is from Anaximander of Miletos and his "Noös": the most pure and subtle of all things, knowledge about all things and infinite power. Thus Anaximander is close to God, one step away from pure theism.

The noösphere must now be redefined. It was originally tightly intertwined with notions of progressive evolution. Once disentangled, the noösphere becomes both a geological phenomenon and a biological/spiritual one. As Martin put it, "life was raised above the phenomena of the inanimate world precisely by feeling." We now see not only life, but also thought as a geological force, neither for "good" nor "evil," but manifest as changes in the quality of interactions between organisms and their environments. The Ediacarans were the first to undertake these developmental steps.

This provides the best explanation for the Cambrian explosion: Brainy predators induce rapid evolution. Martin is wrong when he argues that when "any form of animal life is subjected too intensely to the struggle to survive we find narrow physical adaptations and rigid instinctive reactions which make further development impossible." Sometimes the induced evolution is indeed "regressive," toward small size or confining exoskeletons to avoid predators. Other directions are more supportive of size and complexity, as understanding of symbiogenesis has taught us. The struggles in these directions were no less intense.

Once begun in the Cambrian with the genesis of the metazoan noösphere, the genesis of complex organisms was virtually unstoppable. The development of humanlike consciousness was an inevitable result and could have happened sooner that it did. When humans die out, something very like us could very well reappear. Martin is right when he argues that there is no obvious reason to believe that the process of progressive evolution cannot continue.

The Ediacarans underscore the validity of the view presented above. Here we have organisms, descended from microbes that are also the ancestors of animals. The Lipalian organisms expanded rapidly in size, forming body forms the world had never seen. Most of these creatures were exercising the prerogative of the pressure of life, expanding into new habitats, with a unique cell family approach to multicellularity and an intimacy to their silty and sandy substrate that was so close that in some cases it was difficult to tell where the Ediacaran ended and the sediment began. Life expanded fruitfully in the Lipalian, and both the Ediacarans and the metazoan animals began to use sense organs to collect information. Both then dedicated an organ (the brain) to storing information about the environment, about the *umwelt*. Like a catch basin in the flow of information between organism and environment, memory was born and the world was created anew.

Notes

1. See p. 296 in M. L. West, "*Ab Ovo*: Orpheus, Sanchuniathon, and the Origins of the Ionian World Model," *Classical Quarterly* 44 (1994):289–307.

2. J. Achenbach, "The Final Frontier?" *The Washington Post*, May 30, 1996:C1-C2.

3. "Every mind is its own world"; see p. 110 in E. D. Allen, L. A. Sandstedt, B. Wegmann, and M. M. Gill, *¿Habla Español? Essentials Edition* (New York: Holt, Rinehart and Winston, 1978).

4. G. Scherz, *Steno: Geological Papers* [papers in Latin with parallel English text] (Odense, Denmark: Odense University Press, 1969). In some ways Steno's grappling with the problem of fossilization (that is, his study of shark teeth as fossils and the problem of "a solid within a solid") is recalled by the Ediacarans, these supposedly soft-bodied organisms often preserved in, or even filled with, coarse sand.

5. S. J. Gould, "The Evolution of Life on Earth," *Scientific American* 271 (1994):84–91.

6. cf. Francisco J. Ayala, "Ascent by Natural Selection," *Science* 275 (1997):495–496.

7. See p. 125 in R. Gore, "The Cambrian Period Explosion of Life," *National Geographic* 184, no. 4 (1993):120–135; S. Conway Morris, "Rerunning the Tape," *Times Literary Supplement*, December 12, 1991:6.

8. O. H. Schindewolf, *Der Zeitfactor in Geologie und Paläontologie* (Stuttgart: E. Schweizerbart, 1950).

9. A. Seilacher, "Der Beginn des Kambriums als biologische Wende," *Neues Jahrbuch für Geologie und Paläontology, Abhandlungen* 108 (1956):155–180.

10. Page 7 in D. I. Axelrod, "Early Cambrian Marine Fauna," *Science* 128 (1958):7–9.

11. See pp. xxii–xxiii in L. N. Khakhina, *Concepts of Symbiogenesis: A Historical and Critical Study of the Research of Russian Botanists*, edited by Lynn Margulis and Mark McMenamin (New Haven: Yale University Press, 1992).

12. "Progressive evolutionary change" is used here in the sense of complexification and encephalization; see M. McMenamin and L. Margulis, "Note on Translation and Transliteration, pp. xxix-xxx in L. N. Khakhina, 1992. Francisco J. Ayala ("Ascent by Natural Selection," *Science* 275 [1997]:495–496) raises the important point of what should constitute evolutionary progress. Is it greater complexity, or diversity, or biochemical proficiency? I see evolutionary progress as the steps undertaken via the agency of natural selection that lead to the evolution of life forms with increasing abilities to process, store, and manipulate information about their environment—in other words, the steps that lead to generation of a noösphere.

13. Iterative evolution is re-evolution of particular body forms or behaviors in a special type of repetitive ecologic replacement, in which a "basic parent stock has

given rise to successive groups of higher taxa, each replacing the former." See p. 365 in D. M. Raup and S. M. Stanley, *Principles of Paleontology* (San Francisco: W.H. Freeman, 1978).

14. Nerve cells or their equivalents in Ediacarans may have differed greatly from those in metazoa.

15. See p. 170 in M. F. Glaessner, "A New Genus of Late Precambrian Polychaete Worms from South Australia," *Transactions of the Royal Society of South Australia* 100 (1976):169–170.

16. See p. 190 in M. A. S. McMenamin and D. L. S. McMenamin, *Hypersea: Life on Land* (New York: Columbia University Press, 1994).

17. See pp. 145 and 147 in H. R. Maturana and F. J. Varela, *The Tree of Knowledge: The Biological Roots of Human Understanding* (Boston: Shambhala, 1987).

18. Analogous to, say, the stomata of a plant leaf.

19. Assuming some limited amount of motility.

20. Which could reradiate light to more photosynthetically useful wavelengths.

21. Such organs of sight would not be recognizable on a fossil of *Rimicaris;* C. L. Van Dover, E. Z. Szuts, S. C. Chamberlain, and J. R. Cann, "A Novel Eye in Eyeless Shrimp from Hydrothermal Vents of the Mid-Atlantic Ridge," *Nature* 337 (1989):458–460.

22. V. I. Vernadsky, *The Biosphere*, translated by D. Langmuir, revised by M. McMenamin (New York: Nevraumont/Copernicus, 1997).

23. Including *Vendia, Pseudovendia, Vendomia, Marywadea*, and a *Marywadea*-like form (plate 4, figure 1 of J. G. Gehling, "The Case for Ediacaran Fossil Roots to the Metazoan Tree," *Geological Society of India Memoir* 20 [1991]:181–224), *Bomakellia, Spriggina, Mialsemia*, and "soft-bodied trilobite." *Parvancorina* might belong to this list. In all of these forms the early cell families fuse to form the cephalized region. It is interesting to note that all the cephalo-Ediacarans are late-appearing forms. The interpretation of the "head" as a holdfast is not tenable because its size in the smaller forms (such as *Vendia*) is far beyond what would be required for attachment.

24. See note 2, chapter 11.

25. For example, they might have been able to develop lethal electrical potentials, as are known in a variety of modern electric fishes (eels and rays).

26. A. Seilacher, "An-aktualistiches Wattenmeer," *Paläontologische Zeitschrift* 31 (1957):198–206.

27. M. A. Fedonkin, "Drevneishie iskopaemye sledy i puti evoluutsii povedeniya gryntoedov [Ancient Trace Fossils and the Behavioral Evolution of Mud-Eaters]," *Paleontologicheskii Zhurnal* 2 (1978):106–111.

28. Despite complaints from adherents to the whimper hypothesis (R. A. Fortey, D. E. G. Briggs, and M. A. Wills, "The Cambrian Evolutionary 'Explosion': Decoupling Cladogenesis from Morphological Disparity," *Biological Journal of the Linnean Society* 57 [1996]:13–33), the fact remains that fossils of no more than five metazoan phyla have been found in rocks deposited before the Cambrian; see M. A. S. McMenamin and D. L. S. McMenamin, *The Emergence of Animals: The Cambrian Breakthrough* (New York: Columbia University Press,

1990). These would include cnidaria, possibly annelida, and several other worm-like phyla (as tracemakers), and possibly arthropoda. No one knows how many phyla were extant in the Cambrian, but the higher estimates have ranged from 60 (J. W. Valentine and D. H. Erwin, "Interpreting Great Developmental Experiments: The Fossil Record," in R. A. Raff and E. C. Raff, eds., *Development as an Evolutionary Process*, pp. 71–107 [New York: Alan R. Liss, 1987]) to 100 (McMenamin and McMenamin, 1990). Approximately 37 live today; the new animal phyla Loricifera and Cycliophora have been added in recent years (S. Conway Morris, "A New Phylum from the Lobster's Lips," *Nature* 378 [1995]:661–662; R. Lewin, "New Phylum Discovered, Named," *Science* 222 [1983]:149; J. A. Miller, "Microscopic Animal in a New Phylum," *Science News* 124 [1983]:229). Considering the frequency of severe mass extinctions since the Cambrian, it seems likely that Earth's biota lost a number of phyla during this interval. Most of the phyla lost were probably small phyla, such as Loricifera, represented by few different types of species.

29. Seilacher himself seems to equivocate on this point; see L. W. Buss and A. Seilacher, "The Phylum Vendobionta: A Sister Group of the Eumetazoa," *Paleobiology* 20 (1994):1–4.

30. H. Martin, *The Sheltering Desert*. Translated by W. Kimber (Cape Town, South Africa: A. Donker, 1983).

31. Their vehicle was a 1939 Chevy three-quarter-ton pickup. Chevrolet makes reliable field vehicles, but Nama farmers of the 1930s preferred Ford trucks because their greater maximum speed (70 miles per hour versus 60 miles per hour for the Chevys) gave a better chance of clearing sand dunes (Eldridge M. Moores, personal communication, March 2, 1997). Recall that we used a Chevy Suburban during the Sonoran expedition.

32. H. Martin, *The Sheltering Desert*. Translated by W. Kimber (Cape Town, South Africa: A. Donker, 1983), pp. 230–232.

13 · Revenge of the Mole Rats

The equation of evolution with progress has veered widely from
being self-evident in the eyes of nineteenth-century evolutionists
to being widely regarded as an anthropomorphic delusion in more
recent times.

—Rudolf A. Raff[1]

What constitutes animality . . . is the faculty of utilizing a releasing
mechanism for the conversion of as much stored-up potential
energy as possible into "explosive" actions.

—Henri Bergson[2]

For this final chapter we must depart Germany and German thought
and turn to France.

In 1896, Albert Gaudry, professor of paleontology of the Museum of
Natural History in Paris, published *Essai de paléontologie philosophique:
Ouvrage faisant suite aux enchaînements du monde animal dans les temps
géologiques* (*Essay on Philosophical Paleontology: A Work Made to Follow
the Sequence of the Animal World in Geologic Time*).[3] This book, as indicated in its subtitle, is a sequel to Gaudry's earlier three-volume work on
the sequence of the animal world in geologic time.[4]

Although Gaudry was a well-respected scientist in his day, his books
have been all but forgotten after the passage of more than a century. In
these books, Gaudry takes the first halting steps toward our understanding of the Cambrian explosion and the fall of the Garden of Ediacara.

In *Enchaînements*, written in 1883, Gaudry presents examples of all
of the main fossil types of the Paleozoic and provides commentary on the
sequence of forms displayed in the Paleozoic, fossil record. In the section
on fish (pp. 232–233), Gaudry shows the reader examples of the protective armor of Paleozoic fish. In the book's summary he draws some
conclusions regarding the history of life. But he saves most of his philosophical speculations for the sequel, *Essai de Paléontologie Philosophique*.

From the summary of *Enchaînements*:

The majority of animals found in the Paleozoic, and especially in
the Silurian, appear to have been better suited to defense than to

attack, as if, in the early days of the world, these creatures (which are rare today) had had a greater need to be protected. Thus certain rugosan corals had opercula,[5] [figure 13.1] cystoids were ensconced in skeletal boxes, and similarly disposed were most Paleozoic crinoids, which instead of having their viscera exposed as in Mesozoic crinoids, have them enveloped in a box that recalls that of the cystoids; the brachiopods were only weakly able to open their shells[6]; *Maclurites* and many other Paleozoic gastropods had an operculum; with cephalopods, the aperture of the shell was often constricted. I have noted that when comparing Paleozoic examples to their modern analogs, the ancient prosobranch mollusks were not carnivorous. If, in place of flimsy creatures, protected by a shell or a carapace, hiding themselves in the sediments which formed Paleozoic sedimentary rocks, there instead had been initially those creatures more capable of attack than defense, it is possible that life never would have developed on our planet, and we might have ended up with absolutely nothing instead of ending up with what we do have, the fecund and diversifying expansion of life?[7]

This is an interesting line of thought, but it admits of some logical difficulties. First, is it possible that hypothetical voracious early predators could have wiped out all animal life? I don't think so. Second, if the flimsy and weak animals of the early Paleozoic, barely strong enough to peer out of their shells, had to be protected by heavy skeletons, what were the skeletons protecting them against? Ordinary physical forces such as current and shifting sand? If so, then the protection was grandly overdesigned, as much so as in the proverbial brick outhouse.

Fig. 13. — *Calceola sandalina*, grandeur naturelle; on voit à gauche une valve inférieure, au milieu la petite valve supérieure, et à droite un échantillon avec ses deux valves. — Dévonien de Gerolstein, Eifel. (Collection d'Orbigny.)

FIGURE 13.1: *Calceola sandalina*, a Devonian coral from France distinguished by the fact that it had an operculum or "lid." Width of coral approximately 3 cm.

From p. 15 of A. Gaudry, *Essai de paléontologie philosophique: Ouvrage faisant suite aux enchaînements du monde animal dans les temps géologiques* (Paris: Masson et Cie., Libraires de l'Académie de Médecine, 1896).

In *Essai de paléontologie philosophique*, Gaudry hastens to spell out his philosophical leanings favoring a progressive view of evolution:

> The history of the animal world, considered in the fullness of geologic time, is rather similar to the comparatively short history of man. We see successively:
>> The multiplication of beings on the surface of the globe.
>> Their differentiation.
>> Their enlargement.
>> The elaboration of their behavior.
>> The improvement of their sensory apparatus.
>> The advancement of their intelligence.[8]

The next page reiterates a point made in his earlier book. Gaudry states that "I will say at the outset that the multiplication of creatures has been facilitated because the first arrivals were better defended and less attacked than their descendants."[9] The logic here is still difficult to follow; early animals are not being attacked, yet they require massive protection. If there had been predators at the outset, they might very well have driven everyone to extinction! However, Gaudry is beginning the process of addressing an important question that will shortly be taken up by others: What is the relationship between the origin of skeletons and their use for protection?

Gaudry develops, in an evolutionary sense, a clearer understanding of this question when he discusses the fossil fish:

> With these animals as well, the offensive arms are augmented at the same time as the defensive arms.[10]

He goes on to argue that robust teeth, needed to crush robust armor, diminish in the fossil record as soon as the majority of fish begin to rely more on speed than on armor for protection. Here, Gaudry develops the notion of evolutionary escalation,[11] the idea that prey defenses and predator capabilities are evolutionarily as well as ecologically linked.

Gaudry's work points the way to a deeper understanding of the relationships between predation and prey innovation, and it includes some tantalizing comments regarding the relationship between skeletons and the emergence of animals. Gaudry's work lays the foundation for further thought on the matter.

The next advance in thinking on this subject comes from an unexpected quarter. It is somewhat scandalous for the paleontological profession that it comes not from a paleontologist[12] but from a philosopher who read and cited Gaudry's work. The philosopher's name is Henri

Bergson, and to my knowledge this book is the first paleontological work to acknowledge his priority in this matter.

Henri Bergson is remembered as an influential and highly controversial French philosopher. Bergson was descended from Polish Jews of Warsaw. His grandfather, Berek son of Samuel, had three sons, or Bereksons, hence the name Bergson.[13]

Bergson is best known for his concept of élan vital, the vital force or vital impulse present in matter that drives evolution to more perfect states. Bergson felt that "ultimate reality is a 'vital impulse' and that this reality can be grasped only by metaphysical intuition."[14] Some have viewed Bergson's philosophy as antiscientific and as "an attack on the competence of the intellect. Bergson's object is to vindicate the spiritual principle in nature and to assert the superior competence of intuition."[15] Offended by the implications of his philosophy, secularist intellectuals of France regarded Bergson's philosophy as "pathetic."[16]

Thus it is more than a little ironic that Bergson, on reading Gaudry, becomes the first to correctly understand the relationship between predation and the origin of animals with skeletons. His astonishing insight on this subject is developed in his best-known book, *Creative Evolution*.[17]

Documents are lacking to reconstruct this history in detail, but we can make out its main lines. We have already said that animals and vegetables must have separated soon from their common stock, the vegetable falling asleep in immobility, the animal, on the contrary, becoming more and more awake and marching on to the conquest of a nervous system. Probably the effort of the animal kingdom resulted in creating organisms still very simple, but endowed with a certain freedom of action, and, above all, with a shape so indeterminate that it could lend itself to any future elaboration. These animals may have resembled some of our worms, but with this difference, however, that the worms living today, to which they could be compared, are but the empty and fixed examples of infinitely plastic forms,[18] pregnant with an unlimited future, the common stock of the echinoderms, molluscs, arthropods and vertebrates.

One danger lay in wait for them, one obstacle which might have stopped the soaring coarse of animal life. There is one peculiarity with which we cannot help being struck when glancing over the fauna of primitive times, namely, the imprisonment of the animal in a more or less solid sheath, which must have obstructed and often even paralyzed its movements.[19] The molluscs of that time had a shell more universally than those of today.[20] The arthropods in general were provided with a carapace; most of them were crustaceans.[21] The more ancient fishes had a bony sheath of extreme hardness.[22] The explanation of this general fact should be sought, we believe, in a tendency of soft organisms to defend themselves against one another by making themselves, as far as possible, undevourable. Each species, in the act by which it comes into being, trends towards that which is most expedient. Just as among primitive organisms there were some that turned toward animal life by refusing to manufacture organic out of inorganic material and taking organic substances ready made from organisms that had turned toward the vegetative life,[23] so, among the animal species themselves, many contrived to live at the expense of other animals. For an organism that is animal, that is to say mobile, can avail itself of its mobility to go in search of defenseless animals, and feed on them quite as well as on vegetables. So, the more species became mobile, the more they became voracious and dangerous to one another. Hence a sudden arrest of the entire animal world in its progress towards higher and higher mobility; for the hard and calcareous skin of the echinoderm, the shell of the mollusc, the carapace of the crustacean and the ganoid breast-plate of the ancient fishes probably all originated in a common effort of the animal species to protect themselves against hostile species. But this breast-plate, behind which the animal took shelter, constrained it in its movements and sometimes fixed it in one place. If the vegetable renounced consciousness in wrapping itself in a cellulose membrane, the animal that shut itself up in a citadel[24] or in armor condemned itself to a partial slumber. In this torpor the echinoderms and even the molluscs live today. Probably arthropods and vertebrates were threatened with it too. They escaped, however, and to this fortunate circumstance is due the expansion of the highest forms of life.

In two directions, in fact, we see the impulse of life to movement getting the upper hand again. The fishes exchanged their ganoid breast-plate for scales. Long before that,[25] the insects had appeared, also disencumbered of the breast-plate that had protected their ancestors. Both supplemented the insufficiency of their protective covering by an agility that enabled them to escape their enemies, and also to assume the offensive, to choose the place and the moment of encounter. We see a progress of the same kind in the evolution of human armaments. The first impulse is to seek shelter; the second, which is the better, is to become as supple as possible for flight and above all for attack—attack being the most effective means of defense.[26] So the heavy hoplite was supplanted by the legionary; the knight, clad in armor, had to give place to the light free-moving infantryman[27]; and in a general way, in the evolution of life, just as in the evolution of human societies and of individual destinies, the greatest successes have been for those who have accepted the heaviest risks.

Several years after the publication of Bergson's *Creative Evolution*, paleontologists finally converged on the same idea regarding the relationship between early predators and skeletons. In 1910, a lecturer at Birkbeck College, London, named John W. Evans published an article titled "The Sudden Appearance of the Cambrian Fauna."[28] Apparently unaware of Bergson's writing, which he does not cite, Evans wrote that the "fact that these hard structures [early skeletons] are almost invariably external, indicates that the purpose they served was the protection of the organism from injury" (p. 545). He then contemplates what might make these organisms prone to injury and then infers predators as the likely agent. Evans then considers the problem of the origin of predators:

If it be asked, why no predaceous type had appeared until this comparatively late period of the evolution of animal types, it may be answered that the active carnivorous forms usually belong to the more highly organized groups which are living at any period in the same environment, and as we go backwards in geological times these predaceous types become relatively fewer and less effective, so that it need not surprise us that at a period when the representatives of all the main divisions of the animal kingdom were at a comparatively primitive stage, there should have been no forms of life that preyed actively upon others. [pp. 545–546]

Evans thus saw predators as an "advanced" form of animal life. His logic here is solid, although it now must be noted that the largest of the early Cambrian predators such as *Anomalocaris* (known only from incomplete and misunderstood fossils in Evans's day) appear to be on a par with modern predators in terms of raptorial efficiency if one judges from their morphological traits such as grasping appendages and swimming fins.[29]

Shortly before his death, Bergson wrote that "my reflections have led me closer and closer to Catholicism, in which I see the complete fulfillment of Judaism. I would have become a convert had I not seen in preparation for years a formidable wave of anti-Semitism which is to break upon the world."[30] So it is appropriate that the scientist most influenced by Bergson's thought in this century was the paleontologist and Jesuit priest Pierre Teilhard de Chardin.

Teilhard is best known for his religious concept of the noösphere, the "thinking layer" of the earth that, fitfully at first but with increasing intensity as the lineages of history begin to converge, is in the process of bringing about the fulfillment of creation by attainment of the divine Omega Point. In Teilhard's best-known book, *The Phenomenon of Man*,[31] our species is seen as *the* critical evolutionary step in the attainment of the noösphere and the Omega Point. In the seminal statement of what is now called the New Age movement, Teilhard wrote that the noösphere is "the beginning of a new age. The earth 'gets a new skin.' Better still, it finds its soul" (pp. 182–183). For Teilhard, noögenesis, or development of the noösphere, "rises upwards in us and through us unceasingly" (p. 287). Teilhard's philosophy is unabashedly teleological, indeed, the Omega Point is presented by Teilhard as the end point of evolution.

The concept of the noösphere was developed in Paris at the Sorbonne by Teilhard and Bergson's devoted disciple Édouard LeRoy, after listening to geochemical lectures by materialistic Russian biogeochemist Vladimir Vernadsky. Many scholars credit LeRoy with invention of the term *noösphere*,[32] although Teilhard claimed to have coined it: "I believe, so far as one can ever tell, that the world 'noösphere' was my invention; but it was he [LeRoy] who launched it."[33] Vernadsky was not outspoken about the noösphere but emphasized it in his final paper, published in English shortly before his death.[34] Vernadsky, the first scientist to fully conceive of life as a geological force, extended this view in his final paper to human thought.[35] Impressed by the destructive energies released by two world wars, Vernadsky viewed the noösphere as a new and potent geological force.

Vernadsky was apparently acquainted with Bergson,[36] although there is no record of Bergson and Teilhard ever having met. Teilhard was forbidden to publish his famous book during his lifetime by his ecclesiastical superiors. After his death, the posthumous *Phenomenon of Man*[37] quickly became a classic of popular theology and subsequently was savaged by secular scientists, who were perhaps encouraged by the fact that Teilhard was no longer around to defend his ideas.[38] This is only to be expected for someone with the courage to attempt to cross no-man's land, the divide between the secular and the sacred. Like the French mountain men of the American frontier, Teilhard was an explorer of human thought, searching for a "northwest passage" that would link the mystical, sacred vision of the world with a paleontological understanding of evolution.

The most important, and ironically one of the least known, of the evolutionary debates of the twentieth century occurred between Teilhard and George Gaylord Simpson[39] in 1949 at the Paris *Colloquium sur paléontologie et transformisme* (*Colloquium on Paleontology and Transformism*). Here, in a unique intellectual clash, Teilhard and adherents to the modern synthesis (the neo-darwinian, orthodox view of evolution) tested one another at a formal conference. This event is the only time when Teilhard directly addressed his harshest critics and thus looms in importance considering the posthumous publication of Teilhard's most famous work.

Teilhard began the debate with a paper discussing cases of iterative evolution in fossil Chinese mole rats. Teilhard's research into the evolution of these mole rats,[40] members of the family Siphneidae, revealed that the main trunk of the siphneidid family tree diverged into three separate branches that followed independent evolutionary trajectories.[41] Similar traits then appeared in all three lineages. First, all rats experienced an increase in size. Second, each lineage developed continuous molar growth. Third, all the rats evolved fusion of the cervical vertebrae. Teilhard argued that the example of the Chinese mole rats demonstrated directionality in evolution.

Implicit in this finding, and Teilhard may have made it explicit in his talk, is the likelihood that other bodily features could evolve in similar ways in unrelated creatures. Progressive increase in brain size, seen throughout the history of terrestrial vertebrates, was the type of evolutionary directionality of most interest to Teilhard. For this is the evolutionary direction that leads to the Omega Point.

Simpson, and his neo-darwinian ally T. S. Westoll, countered Teilhard's claims about the siphneidid rats by arguing that this and other cases "of parallel evolution could adequately be explained with the Neo-darwinian concepts of orthoselection."[42] It is not surprising that iterative evolution was Teilhard's favorite line of research, for he realized it exposed a serious weakness in the neo-darwinian synthesis. In his review of Simpson's book (translated from English) *Rythme et modalité de l'évolution*, Teilhard wrote,

During his long career, Dr. Simpson stuck to his position, an intransigent neo-Darwinist attitude, as known to his friends. If one listened to him, everything in zoological evolution should be explicable by the play of selected chances *alone*. Besides the incontestable advantages of this attitude (which obliges the biologist to analyze and to take apart the mechanisms of morphogenesis), it has, we repeat once more, an evident weakness. In its obstinate refusal to look at the indisputable psychic ascent (invention) that globally accompanies the expansion and the arrangement of the biosphere, it deprives the evolutionary process of all direction and all significance as a whole, bringing about the particularly serious result of leaving the human phenomenon unexplained, and scientifically not understandable.[43]

This is a remarkable passage, most remarkable for what it tells us about Teilhard. Teilhard reveals himself as no misty-eyed mystic, but as a scientist willing to stare directly into the face of that which is most threatening to his own view of life. In the Simpsonian view, there is nothing special about *Homo sapiens*. We are just another mammalian species (a member of a dwindling family, at that). Simpson's most ardent living disciples expand on this thought, extending it to a dismal worldview in which evolution has no direction and no purpose. Apparent directionality, in their view, is merely an outcome of the requirement of evolution to go from simple to complex, because it is simply impossible for life to get much less complex than a single cell or virus.

It is clear, then, that Teilhard was ready and willing to confront his critics. But how forceful are his counterarguments? Examples of convergent evolution and iterative evolution, not much researched these days because of their association with Teilhard,[44] deserve renewed and serious evaluation.

Aimée L. MacEachran (a former student of mine) has preliminary results showing simultaneous iterative evolution in two unrelated foraminifera[45] lineages with different skeletal compositions.[46] This result challenges the view that iterative evolution results from pressures of natural selection (orthoselection) on closely related forms. Simpson's riposte to Teilhard in the case of the Chinese mole rats was based heavily on the fact that the mole rat lineages were, in a geological sense, newly separated, and would still have had quite similar genetic heritages. But this close genetic similarity may not hold for other examples of convergent evolution and deserves to be rigorously examined.

An important test would be whether encephalization could have evolved in separate kingdoms, as proposed above for both animals and Ediacarans. It is fitting that this test take place at such an evolutionary starting point. For as Teilhard wrote in the Russian edition of *The Phenomenon of Man*,

> Both in time and in space the growing tip of a phyletic branch has a minimum of differentiation, a minimal force of expansion, and a minimum resistance and this is why the early stages of evolution occur quickly. What influence does time have on this frail and pliable bud?[47]

Teilhard's answer, of course, is that the bud will follow certain trajectories governed by evolutionary laws requiring an increase in evolutionary complexification.[48] Many buds will follow similar trajectories (convergent evolution) or repeated trajectories (iterative evolution). For Teilhard, the direction is always in the direction of Omega, especially when the branches begin to converge.

Here is the profound gap between Teilhard's thought and Simpson's. For Simpson, chance is all; purpose does not exist. For Teilhard, chance exists but is subordinate to the directional forces in evolution. There is tremendous distance between these two intellectual styles, and the tension between them has not been adequately resolved.[49]

Considering that Teilhard studied fossil mole rats, it is interesting to note that modern mole rats are playing a key role in the continuing evolutionary debate. In 1981 a zoologist reported an instance of eusociality[50] in the naked mole rat *Heterocephalus glaber* (family Bathyergidae).[51] This was an astonishing discovery. Eusociality, well known in insects such as bees and ants, was heretofore unknown in any type of vertebrate, let alone a mammal. As the genus name *Heterocephalus* might seem to suggest, these little creatures were "differently minded" in comparison to other mammals when it came to behavioral evolution.

Naked mole rats are, quite frankly, repulsive little creatures known from the arid regions of Kenya, Somalia, and Ethiopia. They have been called sausages with fangs, and Rudolf A. Raff calls their species the "downright most desperately ugly mammal on Earth."[52] They inhabit extensive underground foraging tunnels, where they dig for tubers and eat each other's feces. They are the only subterranean mammals to live in large colonies, and as in eusocial insects, only a single female, the queen, breeds. Other comparisons with social insects (especially termites) include long life spans, male and female members belonging to several different working castes with different duties, overlap of generations, cooperative brood care, and queen activation (by pushing and shoving) of lazy workers.[53]

The discovery of the eusocial mole rats was considered a decisive victory, a Revenge of the Mole Rats, by some neo-darwinian/Simpsonian evolutionists. Here, they felt, was an example of a species within the supposedly exalted class Mammalia. But instead of making the intrepid climb toward Omega, these mammals were scrabbling underground, dronelike, in what strongly resembled a colony of insects. In the language of sociobiology (itself an outgrowth of the neo-darwinian modern synthesis), the naked mole rats were slaves to their genomic heritage, and their curious behavior was rigidly constrained by genetic programming. On the face of it, it appeared to be the ultimate vindication of the doctrine of the selfish gene, and a final and fatal insult to Teilhard's legacy of evolutionary thought.

But the interpretation of naked mole rats cannot be so easily and neatly resolved in favor of the neo-darwinians. One large difference between the eusocial mole rats and the eusocial insects begged for an explanation. As noted in the groundbreaking 1981 paper, "*Heterocephalus* differs from the eusocial insects in not having a clearly defined reproductive male" or male dispersal morph capable of seeking out new colonies and relieving the inbreeding endemic to naked mole rat colonies.[54] But in 1996, M. Justin O'Riain and his coauthors reported discovery of a male dispersal morph in the naked mole rat species *Heterocephalus glaber*.[55] Thus the comparison with eusocial insects is complete and completes a most astonishing instance of convergent evolution.

Teilhardians take heart in this result and could call it the Second Revenge of the Mole Rats. For if evolution can evoke such profound behavioral convergences in such dramatically unrelated animals as termites and naked mole rats, then a large amount of what Teilhard has written about directionality in evolution gains credibility. There are forces afoot that dramatically constrain the course of evolutionary change. Simpson's complaint about directed evolution in the fossil mole rats, a criticism based on the fact that they shared a close genetic heritage and thus would have been expected to respond similarly to similar selective pressures, totally and unexpectedly collapses when confronted with the case of the modern eusocial mammals. Convergent evolution and iterative evolution demonstrate that evolution can be directional and is quite capable of "going somewhere."

Whether or not it is indeed doing so remains to be seen. One could look at the eusocial mole rats as a dead end, as Henno Martin put it, slave to "rigid instinctive reactions which make further development impossible." The mole rats themselves, meanwhile, are more intelligent than average mammals of their size and live their obscure lives in what might be interpreted as a happy state, if one enjoys selfless service to others. Animal behaviorists have recently discovered yet another remarkable trait in these gerbil-size rodents: selfless altruism. When a foraging mole rat finds an easily transported piece of tuber or root, "it scurries home with the food, chirping all the way."[56] Returning to the colony, the mole rat continues to chirp while waving the food around, alerting other members to the presence of more food. The scout then heads out again with a troop of helpers to further provision the colony. Only after the colony is sated do the foragers eat.

If mole rats could contemplate evolutionary theory and speak about it, what would they say? I think they might have disparaging things to say about the pessimistic and self-indulgent ideas of the neo-darwinians. We might even hear some reactionary comments about the *human* constitution (as in, "Who are you calling downright desperately ugly?"). This would be the Third and Final Revenge of the Mole Rats.

Is evolution going anywhere? The late poet laureate of the United States, Joseph Brodsky, addressed this question in his poem *The Butterfly*. Brodsky suggested that beauty of the butterfly is so fleeting as to challenge the idea that the world has an end goal or *telos*. For why would such a glorious creature be created for such an ephemeral lifespan? Brodsky leaves open the possibility that there is indeed some sort of universal *telos*, but completes the poem with a unique variation on this theme: "and if as some would tell us/there is a goal,/it's not ourselves." Or is it?

Here is the reason I have called this book *The Garden of Ediacara*. The Garden of Eden religious story and the scientific analysis of the Ediacarans form an inseparable pair, a diadic coupling that must remain intact in order for science (construed as real, if incomplete, knowledge about things) to thrive. If we try to purge science of the irrational and to banish intuitive approaches in science, as many orthodox neo-darwinists have tried to do, we risk destroying science.

Teilhard's work has merited renewed scrutiny by at least one evolutionist in recent years. Chet Raymo, who teaches science at Stonehill College and writes a science column for the *Boston Globe*, writes that Teilhard's effort to reconcile science and religion struck Raymo at first like "a jolt of electricity," and induced in him something akin to a "hallucinogenic" vision. Years later, Raymo returned to Teilhard's work, only to find the theological statement of faith of a "God-struck dreamer . . .

hopeless as a program for bridging the gulf between science and theology." Teilhard's great mistake, according to Raymo, was insisting that his most famous work be read as a "scientific treatise" rather than a work of theology. For example, it is unclear how Teilhard's concepts of radial energy[57] could be rigorously tested, and such "vagueness disqualifies Teilhard's ideas as science." Raymo concedes, nevertheless, that "Teilhard's vision may yet turn out to be correct." Trendy concepts promoted by neo-darwinists such as self-organization may be evidence of a natural drive toward "complexity and perhaps consciousness." And with the Internet and the World Wide Web wrapping the world in a noöspheric embrace, and physicists also taking a second look at Teilhard's cosmology, there is a growing sympathy for Teilhard's vision in the contemporary scientific community.[58]

In the final analysis, Bergson must be right. Intuition trumps reason. From Bergson to *The Urantia Book*, the human mind in all its unpredictable glory is still our most potent scientific tool. As Paul Feyerabend pointed out in 1988, "Modern science survived only because reason was frequently overruled."[59]

The concept of the vital force was not original with Bergson. Airstotle used the word *entelechy* to describe the "realization of potential." Augustine believed that "material things belong to the lowest level of being. Within them God has implanted certain *rationes seminales*; because of these 'seed-like causes' or potencies, new forms appear in the course of time."[60] With evidence in our hands of convergently evolved protective skeletons and eusocial animals, plus numerous cases of iterative evolution, and not only the convergent evolutionary enlargements of brains but perhaps even iterative evolution of the brain itself, we must now accept a neovitalistic view of evolutionary change.

Vitalism and its modern variant, neovitalism, have had rough sledding for most of the twentieth century. A primary sympathetic source on this subject is Hans Adolf Eduard Driesch's 1915 book *The History and Theory of Vitalism* (London: Macmillan and Company). Driesch (1867–1941) was a biologist and philosopher, educated at the universities of Frieburg, Munich, and Jena, who worked from 1891 to 1900 at the Stazione Zoologica of Naples. Driesch's reputation as a scientist is seriously compromised today because of his advocacy of telekinesis and the field of parapsychology; his work on the subject is still in print.[61]

Vitalists like Driesch (he called himself a neovitalist) took comfort from Bergson's *Creative Evolution* (Driesch gives the book a glowing citation in *History and Theory of Vitalism*) but were subjected to blistering criticism by materialists such as Russian geochemist Vernadsky. In 1944, Vernadsky wrote, "New vitalistic notions have their foundation not in scientific data, which are used rather as illustrations, but in philosophical concepts such as Driesch's 'entelechy.' The notion of a peculiar vital energy (W. Ostwald) is likewise connected with philosophical thought rather than with scientific data. Facts do not confirm its real existence."[62] Vernadsky's attack was devastating, and Driesch's advocacy of vitalism proved to be nearly indefensible within a decade. Although Driesch correctly identified the importance of "chance" and "contingency" for the Darwinian scheme, he failed in his flippant critique of Darwinism, relying on a specious argument used today only by creationists: "Darwinism (explains) how by throwing stones one could build houses of a typical style."[63]

But far worse for vitalism was the abject failure of Driesch's "empirical proofs." Driesch presented three pieces of "evidence" favoring vitalism: (1) analytic experimental embryology (*Entwicklungsmechanik*) showed that parts of animals could be regenerated by embryos and adults, evidence of some vague vital force; (2) egg cells undergo division indefinitely and yet remain what they were, with no apparent limit to the repeatability of the process; and (3) mechanistic philosophy cannot explain the actions of man.

"Proofs" 1 and 2 are easily dismissed today with an understanding of the genetic code, which explains both the genetic regulation of regeneration in adults and embryos, and the potentially infinite number of divisions of reproductive DNA. Indeed, the discovery of DNA surprised many scientists who were expecting the secret of the genome to reveal unknown laws of physics and chemistry. The fact that it was a simple chemical trick with no special forces was a terrible blow to vitalism and a boost to mechanistic materialism; in a nontrivial sense, the anticipated "vital force" turned out to be the molecular biology of DNA. "Proof" 3 is more reasonable (recall Teilhard's admonishments about rendering the human phenomenon scientifically understandable in his critique of Simpson) but is rather vague and has been addressed by neo-darwinians in their promotion of the doctrine of sociobiology. Thus Driesch's vitalistic views have become, in the eyes of most scientists, less and less tenable with each passing decade.

Driesch's vitalism was founded on "entelechy," a term he borrowed from Aristotle to denote "the controlling, but immaterial and non-physical, principle of all Life and all organisms."[64] Driesch makes an insightful portrayal of the ethical implications of a materialist, nonvitalistic view of the world,[65] and he blames this dismal worldview on an errant and incomplete rationalism. He suggests that the problem dates back to the origin of animal predators in the Cambrian, and he further makes the remarkable proposals that we must both champion animal-rights vegetarianism and somehow return to the Garden of Ediacara by humanely eliminating all animal predators, effectively turning the clock back on the Bergsonian Cambrian explosion which had[66]:

> created those monsters which we call beasts of prey in the widest sense of the term—those monstrous phenomena where animal turns against animal, to which class we still belong. In the sphere of human soul-life it makes the individual the slave of his impulses and "feelings." By it the individual is rent, and with him spiritual mankind. *Homo homini lupus*[67]—that is, man is a wolf to his neighbor. . . .
>
> . . . It may sound fantastic to intend the abolition of beasts of prey (without cruelty to them, of course); but science knows no star but hope.

Certainly, such proposals add little to Driesch's scientific credibility. Nevertheless, Driesch did develop an interesting classification of the types of teleology applied to biological situations (i.e., goal-directed biological change). He classified teleology as either static teleology or dynamic teleology[68]:

> [Are] the processes of life to be judged teleological only in virtue of their given order, only because a given mechanical form lies beneath them, while every single one is really a pure physical or chemical process [static teleology]—or are the processes of life purposive because of an unanalysible autonomy [dynamic teleology]?

Driesch noted further that static teleology led to a mechanistic theory of organisms, whereas dynamic teleology, with its focus on the "autonomy of vital processes," led to vitalism. Put in these terms, I would have to champion a static version of teleology, apparently the same version to which Henno Martin subscribed, and this is what I am calling here neovitalism. Any "vital forces or energies" are inherent to all matter rather than being anything special or out of the ordinary. As Martin says, the "spiritual force was at work in all feelings,[69] even the simplest." Thus our neovitalism is a structural teleology, an outcome of the structure of the universe and the way it is put together. No mysterious *élan vital* is required in this neovitalism. Directed evolution of life, and the subsequent eventual appearance of intelligent life, is a result of, as German philosopher Kant would put it, *the world as it is.*[70]

The intellectual demise of Driesch's vitalism was largely due to the weakness of his "empirical proofs," which proved to be particularly vulnerable to the onslaught of modern molecular biology in the wake of the discovery of DNA. Like telekinesis, the hoped-for new vital forces of chemistry never materialized as the study of biotic reproduction proceeded to the molecular level.

Now is the time, however, for us to be on guard lest we discard the baby with the bathwater. The neovitalism I am advocating here may be a static teleology in Driesch's scheme, but it will help us to understand how various factors operating at different times and places can constrain and direct evolutionary change.

Please don't misunderstand me; with neovitalism I am not invoking some type of mystical force to accomplish these changes. Rather, there must be something about the structure of the material world that causes matter to organize in this particular and very interesting way. In other words, it would appear that life evokes mind. There is indeed some kind of evolutionary directionality and vital potency.

This is a fully scientific statement, rich with possibilities for analysis, investigation, and generation of new knowledge about our world. For example: Why and how does life develop in this way? To what sort of *telos* (if any) might this process be directed? It is incumbent on us to ask these questions, and if we are able, to learn the answers. The answers will be found in some of the oddest places.

Perhaps the place to begin is with the boojum, a desert plant as odd as any I have seen. Let us return once more to the Sonoran desert.

On August 2, 1976, shortly after we had both graduated from high school, my friend John Kingeter sent me a postcard from Tucson. The postcard[71] carried a color image of a very strange cone-shaped desert plant called the cirio, or boojum.[72] The plant looks like a green baseball bat, without the knob on the end, stuck into the ground (broad end down) with what appear to be cloves stuck into its smooth green surface in a fairly regular pattern. On the text side of the postcard was an account, courtesy Glenton G. Sykes, of how the boojum got its name:

> In 1922 an expedition was organized by Mr. Godfrey Sykes of the Desert Botanical Laboratory at Tucson to study some strange plants found near Puerto Libertad, Sonora, by his son Gilbert. The party reached the area late one afternoon, and Mr. Sykes

focused his telescope on the hills where the plants had been seen. Then, in the words of his son, Glenton, who was also present, "he gazed intently for a few moments and then said, 'Ho, ho, a boojum, definitely a boojum.' " The name took hold then and there and has now become more or less general as the common name. The term "boojum" being taken, of course, after Lewis Carroll's "Hunting of the Snark," a delightful, mythical account of exploration in far-off, unheard-of corners of the world and wherein did abound a legendary creature or thing termed the "boojum," and which was said to dwell upon distant, unfrequented desert shores; hence the name given on the spur of the moment by Godfrey Sykes was perhaps . . . appropriate.

Another strange desert plant from Baja California, the elephant tree, or *torote blanco*,[73] lives in a similar part of the world but belongs to a different family (Anacardiaceae). The boojum and the elephant tree have a similar strategy for surviving in this harsh desert environment. Like many desert plants, they exhibit convergent evolution of succulent water-storing stems. But the stems of the boojum and the elephant tree have another interesting convergence. Both plants have stems that recycle endogenous (internally generated) carbon dioxide. In other words, they refix or recycle respiratory carbon dioxide. This ability to recycle carbon dioxide, which captures and internalizes a tiny part of Earth's carbon cycle, so ensures survival during extreme conditions that both types of plant can survive 5 or more years without rain.[74] Both plants also have nonsucculent leaves to permit high productivity during times of favorable environmental conditions.

Plants of arid regions are greatly varied in their sizes and shapes, but shared solutions to the problems of desert life seem to be the rule rather the exception. From Joshua tree to kokerboomwood, boojum to elephant tree, from paired leaves in *Lithops* to paired leaves in *Welwitschia*, natural selection has guided evolution to similar locations. And from *Marywadea's* "head" to Mary Wade's[75] mind, we find evidence indicating that in many biological situations there may be a limited number of ways of solving certain problems of life. And certain patterns, such as the development of brains, are in an evolutionary sense strongly predisposed to reappear. Before the newfound understanding of naked mole rats, it would have been ridiculous to speculate that insect-style eusociality could occur in mammals. And yet there it is, another compelling pattern. In the jargon of chaos theory, both eusociality and encephalization represent biological attractors.

If the Ediacarans were in the process of developing brains independently of the metazoa, then even the development of something as impressive as the human central nervous system is the result of a common and repeatable evolutionary trajectory. But I don't think that this makes our species any less wonderful or interesting.

I leave you now with a Joshua tree principle of evolution: Once we can see noöspheric development for what it is, we can more fully understand the evolutionary processes involved in our creation, and we may begin to see ourselves for what *we* truly represent.

Notes

1. See p. 98 in R. A. Raff, *The Shape of Life: Genes, Development, and the Evolution of Animal Form* (Chicago: University of Chicago Press, 1996).

2. See p. 120 of H. Bergson, *Creative Evolution* (New York: H. Holt, 1913).

3. A. Gaudry, *Essai de paléontologie philosophique: Ouvrage faisant suite aux enchaînements du monde animal dans les temps géologiques* (Paris: Masson et Cie., Libraires de l'Académie de Médecine, 1896).

4. The first volume concerns us here: A. Gaudry, *Les Enchaînements du monde animal dans les temps géologiques*, vol. 1, *Fossiles primaires* (Paris: Libraire F. Savy, 1883). Gaudry uses the term *primary* in the sense of Paleozoic. The other two volumes are on secondary (Mesozoic) fossils (1890) and tertiary mammals (1895).

Gaudry, highly respected in his day, is today a source of embarrassment for secular French scientists, who scornfully label Gaudry the "spiritual father" of Fr. Pierre Teilhard de Chardin (G. Laurent, "Paléontologie et évolution chez Teilhard de Chardin (1881–1955)," *Mémoires de la Société Géologique de France*, Nouvelle Série 168 (1995):97–100.

5. Gaudry illustrates this interesting "capped" coral in both *Enchaînements* and *Essai de paléontologie philosophique.*

6. This may indeed have been the case; note the absence of muscle scars in the early brachiopod *Mickwitzia* (see M. A. S. McMenamin, "Two New Species of the Cambrian Genus *Mickwitzia*," *Journal of Paleontology* 66 (1992):173–182).

7. Translation mine, pp. 293–294.

8. Translation mine, p. 12.

9. Page 13: "Je dirai d'abord que la multiplication des êtres a été facilitée parce que les premiers arrivés ont été mieux défendus et moins attaqués que leurs descendants."

10. Translation mine, p. 80.

11. An idea developed further, but without attribution to Gaudry, by Geerat J. Vermeij, *Evolution and Escalation* (Princeton, N.J.: Princeton University Press, 1987).

12. In 1894, W. K. Brooks ("The Origin of the Oldest Fossils and the Discovery of the Bottom of the Ocean," *Journal of Geology* 2 [1894]:455–479) implicitly linked the appearance of Cambrian animals with skeletons and the use of skeletons as protection. His thought on the matter, as the title of his article indicates, is in

fact a corollary of his thesis about the colonization of the sea bottom. He sees skeletons as required for protection only after animals colonize the sea bottom, where competitive pressures (not necessarily all from predators) were amplified by the limited space. Thus Brooks sees acquisition of protective skeletons as secondary to the "discovery of the bottom" of the sea. This seafloor discovery was also championed by P. E. Raymond ("Pre-Cambrian Life," *Geological Society of America Bulletin* 46 [1935]:375–392). This idea, premised on the presence of unfossilizable microscopic metazoan ancestors, is difficult to test at best. A. Seilacher ("Der Beginn des Kambriums als biologische Wende," *Neues Jahrbuch für Geologie und Paläontology, Abhandlungen* 108 [1956]:155–180) challenged Brooks by noting (p. 155) "that animals discovered the bottom of the ocean long before the Cambrian fauna appeared." J. William Schopf, Bruce N. Haugh, Ralph E. Molnar, and Donna F. Satterthwait ("On the Development of Metaphytes and Metazoans," *Journal of Paleontology* 47 [1973]:1–9) note the problems with this line of reasoning, noting (p. 6) that "it remains to be suggested what 'caused' these early metazoans to 'discover the bottom.' "

13. B.-A. Scharfstein, *Roots of Bergson's Philosophy* (New York: Columbia University Press, 1943).

14. Page 547 in A. Szathmary, "Bergsonism," pp. 547–550 in *The Encyclopedia Americana*, vol. 3 (New York: Americana Corporation, 1956).

15. Anonymous (probably A. Szathmary), "Creative Evolution," pp. 170–171 in *The Encyclopedia Americana*, vol. 8 (New York: Americana Corporation, 1956).

16. J. Benda, *Une Philosophie pathétique* (Paris: Cahiers de la Quinzaine, 1913).

17. H. Bergson, *L'Evolution créatrice*, 4th ed. (Paris: Librairies Félix Alcan et Guillaumin Réunies, 1908). This translated passage is modified from the 1913 translation (pp. 129–132, which corresponds to pp. 141–144 in the original French) by Arthur Mitchell (*Creative Evolution*, New York: H. Holt).

18. In recent years evolutionary theorists have championed a "plastic" view of early animal genomes and a "constrained" view of later genomes, with consequent loss in the ability of animals to evolve major new innovations; see J. W. Valentine and D. H. Erwin, "Interpreting Great Developmental Experiments: The Fossil Record," in R. A. Raff and E. C. Raff, eds., *Development as an Evolutionary Process*, pp. 71–107 (New York: Alan R. Liss, 1987). Dianna McMenamin and I have criticized this idea, calling it the green genes hypothesis, in M. A. S. McMenamin and D. L. S. McMenamin, *The Emergence of Animals: The Cambrian Breakthrough* (New York: Columbia University Press, 1990). Perhaps in response to our criticism, James W. Valentine now agrees that "there are no indications that the creative ability of regulatory genomes has been impaired" (p. 190 in J. W. Valentine, "Why No New Phyla After the Cambrian? Genome and Ecospace Hypotheses Revisited," *Palaios* 10 [1995]:190–194).

19. Bergson emphasizes here the restrictive nature of early skeletons, whereas Gaudry emphasized their protective function.

20. This is probably true, except that the early mollusk skeletons were in the form of "coat-of-mail-like" multielement scleritomes.

21. Indeed, true crustaceans are known from the Cambrian; D. E. G. Briggs, "The Morphology, Mode of Life, and Affinities of *Canadaspis perfecta* (Crustacea:

Phyllocarida), Middle Cambrian, Burgess Shale, British Columbia," *Philosophical Transactions of the Royal Society of London* B281 (1978):439–487.

22. Bergson cites Gaudry here, and his footnote reads as follows: "See, on these various points, the work of Gaudry: *Essai de paléontologie philosophique*, Paris, 1896, pp. 14–16 and 78–79."

23. Consider here the distinction between heterotrophs and autotrophs.

24. The same simile ("a mediaeval castle") was independently used by C. R. C. Paul, "Early Echinoderm Radiation," in M. R. House, ed., *The Origin of Major Invertebrate Groups*, pp. 415–434 (New York: Academic Press, 1979).

25. Bergson has the timing wrong here; fishes actually appear before insects in the history of life.

26. This sentence anticipates the mathematical analysis presented in M. A. S. McMenamin, "The Cambrian Transition as a Time-Transgressive Ecotone," *Geological Society of America Abstracts with Program* 24 (1992):62.

27. Without directly citing Bergson, Martin F. Glaessner (p. 174 in *The Dawn of Animal Life: A Biohistorical Study* [Cambridge, England: Cambridge University Press, 1985]) ridiculed this line of thought: "The naive assumption that shells are acquired because they protect soft bodies seems influenced by anthropocentric thinking: man uses shields for protection from aggressors."

28. *Congrès Géologique International. Comptes Rendus de la Session* 1 (1910): 543–546.

29. D. Collins, "The 'Evolution' of *Anomalocaris* and Its Classification in the Arthropod Class Dinocarida (nov.) and Order Radiodonta (nov.)," *Journal of Paleontology* 70 (1996):280–293.

30. Scharfstein, 1943, p. 99.

31. P. Teilhard de Chardin, *The Phenomenon of Man* (New York: Harper & Row, 1959). Originally published in French as *Le Phénomène humain* (Paris: Editions du Seuil, 1955).

32. E. LeRoy, *L'Existence idéaliste et le fait de l'évolution* (Paris: Boivin et Cie, 1927). See p. 27.

33. See p. 120 in R. Speaight, *The Life of Teilhard de Chardin* (New York: Harper & Row, 1967).

34. V. I. Vernadsky, "The Biosphere and the Noösphere," *American Scientist* 33 (1945):1–12.

35. V. I. Vernadsky, *The Biosphere*, translated by D. Langmuir, revised by M. McMenamin (New York: Nevraumont/Copernicus, 1997).

36. In the George Vernadsky archives at Columbia University is a photo of Henri Bergson (1859–1941) that may have belonged to his father, Vladimir Vernadsky. Archival listing is as follows: "Photographs are chiefly of George and Nina Vernadsky and their friends and relations." PROD Archival LON; NYCR85-A802; SEARCH CTYV-REF; Record 5 of 14; 17. Photoprints; IX Bergson, Henri, 1859–1941. V. Vernadsky and Bergson met in 1923 (J. Grinevald, personal communication).

37. Teilhard de Chardin, 1959.

38. E. O. Dodson, *The Phenomenon of Man Revisited* (New York: Columbia University Press, 1984).

39. Simpson, perhaps the best known paleontological architect of the neo-darwinian modern synthesis, is today revered by secular evolutionists.

40. P. Teilhard de Chardin, "New Rodents of the Pliocene and Lower Pleistocene of North China," *Publications of the Institut de Géobiologie* 9 (1942):1–101.

41. An evolutionary pattern that resembles some forms of iterative evolution.

42. See p. 161 in L. Galleni, "Relationships Between Scientific Analysis and the World View of Pierre Teilhard de Chardin," *Zygon* 27 (1992):153–166. See also L. Galleni, "How Does the Teilhardian Vision of Evolution Compare with Contemporary Theories?" *Zygon* 30 (1995):25–45.

43. P. Teilhard de Chardin, "Quantitative Zoology According to Dr. G. G. Simpson," *Geobiologia* 1 (1943):139–141.

44. Which is a distinct professional liability among contemporary evolutionists, the overwhelming majority of whom are neo-darwinists.

45. Foraminifera are a type of marine protist with a calcareous skeleton.

46. Aimée L. MacEachran, *Iterative Evolution* sans *Associated Extinction: The Agglutinated and Calcareous Paleozoic Foraminifera*, unpublished honors thesis (South Hadley, Mass.: Mount Holyoke College, 1996).

47. See p. 388 in M. A. Fedonkin, "Vendian Body Fossils and Trace Fossils," in S. Bengtson, ed., *Early Life on Earth*, pp. 370–388 (New York: Columbia University Press, 1994).

48. M. McMenamin and L. Margulis, "Note on Translation and Transliteration," in L. N. Khakhina, *Concepts of Symbiogenesis: A Historical and Critical Study of the Research of Russian Botanists*, edited by Lynn Margulis and Mark McMenamin, pp. xxix–xxx (New Haven: Yale University Press, 1992).

49. Despite the claims of many neo-darwinists.

50. Eusociality is cooperative breeding in a colony.

51. J. V. M. Jarvis, "Eusociality in a Mammal: Cooperative Breeding in Naked Mole-Rat Colonies," *Science* 212 (1981):571–573.

52. Page 368 in R. A. Raff, *The Shape of Life: Genes, Development, and the Evolution of Animal Form* (Chicago: University of Chicago Press, 1996).

53. H. K. Reeve, "Queen Activation of Workers in Colonies of the Eusocial Naked Mole-Rat," *Nature* 358 (1992):147–149.

54. See p. 572 in V. M. Jarvis, "Eusociality in a Mammal: Cooperative Breeding in Naked Mole-Rat Colonies," *Science* 212 (1981):571–573.

55. M. J. O'Riain, J. V. M. Jarvis, and C. G. Faulkes, "A Dispersive Morph in the Naked Mole-Rat," *Nature* 380 (1996):619–621. S. Conway Morris has also noted the "striking convergences" between insects and mole rats ("Trapped in a Hall of Mirrors?" *Trends in Genetics* 12 [1996]:430–431).

56. Anonymous, "Naked Selfless Mole Rats," *Discover* 18, no. 3 (1997):16.

57. According to Raymo, radial energy "drives the material world towards complexity and consciousness." It is thus the engine of Teilhard's process of complexification.

58. C. Raymo, "The Evolution of Belief," *Notre Dame Magazine* Spring (1996):28–31; M. Heller, "Teilhard's Vision of the World and Modern Cosmology," *Zygon* 30 (1995):11–23.

59. See p. 7 in P. Feyerabend, *Against Method* (London: Verso, 1988).

60. See p. 546 of J. K. Ryan, "Augustinianism," in *The Encyclopedia Americana*, vol. 2, pp. 545–547 (New York: Americana Corporation, 1956).

61. H. Driesch, *Psychical Research: The Science of the Super-Normal.* Translated by T. Besterman (North Stratford: Ayer Company Publishers, 1975).

62. Page 509 in V. I. Vernadsky, "Problems of biogeochemistry, II," *Transactions of the Connecticut Academy of Arts and Sciences* 36 (1944):483–517.

63. See p. 137 in *The History and Theory of Vitalism* (London: Macmillan and Company, 1915).

64. W. H. Johnson, "Translator's Note," pp. 11–12 in H. Driesch, *Man and the Universe* (New York: Richard R. Smith, 1930).

65. See pp. 160–163 in H. Driesch, *Man and the Universe* (New York: Richard R. Smith, 1930).

66. See pp. 158–159 in H. Driesch, *Man and the Universe* (New York: Richard R. Smith, 1930).

67. Here Driesch misquotes, as did S. Freud, line 495 of the Roman playwright Plautus's *Poenulus* ("The Little Carthaginian"). The complete line from the play actually refers to the arrival of a stranger:

> lupus est homo homini, non homo
> quom qualis sit non novit
> [Man is no man, but a wolf, to a stranger]

Freud apparently "shortened the saying [to 'lupus est homo homini'] to form an epitome of aggression" (p. ix in P. Bovie, "Preface," pp. vii–xi in D. R. Slavitt and P. Bovie, eds., *Plautus: The Comedies,* Volume III [Baltimore: The Johns Hopkins University Press, 1975]).

68. See p. 5 in *The History and Theory of Vitalism* (London: Macmillan and Company, 1915).

69. Driesch had disparaging things to say about the link between "spiritual forces" and "feelings" emphasized here by Martin. Writing at a time immediately before the economic collapse of 1929 and the rise of Hitler, Driesch worried about this linkage (see p. 117 in *Man and the Universe*), "especially in Germany, where people have long been indulging and still indulge in vague 'feelings.' We may ask, what has been the result for Germany?"

70. Note the similarity of this neovitalism to the best statements of the anthropic principle (J. Polkinghorne, "So Finely Tuned a Universe," *Commonweal,* August 16, 1996, pp. 11–18). Physicists such as Polkinghorne have a long history of careful thought on this and related subjects. Earlier in this century, German physicist Felix Auerbach compared the entropy of inert matter to what he called the "ectropy" of living form, linking it to the evolution and development of life (F. Auerbach, *Ektropismus oder die physikalische Theorie des Lebens*, Leipzig, W. Engelmann, 1910).

71. "Cirio or Boojum: How the Boojum Got Its Name" (card #B5857), published by Petley Studies, 4051 E. Van Buren, Phoenix, AZ 85008, photography by Mike Roberts, Berkeley, CA 94710; G. Sykes, *A Westerly Trend, Being a Veracious*

Chronicle of More Than Sixty Years of Joyous Wanderings, Mainly in Search of Space and Sunshine (Tucson: Arizona Pioneers Historical Society, 1944).

72. The boojum tree or cirio is *Fouquieria columnaris* (Kellogg) Hendrickson, family Fouquieriaceae. See J. Hendrickson, "A Taxonomic Revision of the Fouquieriaceae," *Aliso* 7 (1972):439–537.

73. The elephant tree, or *torote blanco*, is *Pachycormus discolor* (Benth.) Coville, family Anacardiaceae.

74. E. Franco-Vizcaíno, G. Goldstein, and I. P. Ting, "Comparative Gas Exchange of Leaves and Bark in Three Stem Succulents of Baja California, Mexico," *American Journal of Botany* 77 (1990):1272–1278.

75. The paleontologist for whom *Marywadea* is named.

Epilogue: Parallel Evolution

> Mystical thought will expand wherever the foundations of empirically based knowledge are absent or infirm. In evolutionary research, an inappropriately applied holism has caused researchers to neglect the proper approach to scientific analysis of cause and effect.
>
> —Otto H. Schindewolf[1]

Adolf Seilacher's predecessor at the University of Tübingen was professor Otto Schindewolf. Schindewolf was an imposing figure in paleontology (his name means "wolf-skinner"), and he apparently had a tremendous influence on young Dolf Seilacher. It seems fitting that, in 1972, Seilacher was author of his great teacher's obituary.[2]

Perhaps Schindewolf's most influential book was *Grundfragen der Paläontologie* (*Basic Questions in Paleontology*).[3] The final section of this book, titled "Systematics of Parallel Lineages," is devoted to the question of convergent evolution and its implications for the classification of fossil organisms.

Schindewolf did not like to use the term *convergent evolution* because he felt that it implied the notion of an evolutionary "coming together" from separate points. Teilhard de Chardin approved of this aspect of the term, seeing convergence as literal progress toward Omega. Schindewolf usually substituted the term *parallel evolution* in place of *convergence* or *convergent evolution.* Either way you name it, the concept was first defined by Ernst Haeckel.[4]

Schindewolf identified examples of convergence in vertebrates, plants, and numerous shelled cephalopods, especially ammonites. In his discussion of parallel evolution, Schindewolf noted that several related lineages underwent the same evolutionary changes, thus complicating the task of classifying ancient organisms based on their family (phylogenetic) relationships. For example, Schindewolf showed in his figure 4.8 three different lineages of the Devonian cephalopod genus *Cheiloceras* evolving separately into the genus *Sporadoceras.* These three examples of *Sporadoceras,* in turn, independently evolve into the genus *Discoclymenia.* Naturally, this could lead to some confusion in the use of the basic taxonomic term *genus.* Aren't all members of a given genus supposed to be members of a single lineage? Schindewolf acknowledges that this would be the ideal case, but it is not always possible to compose a vertical (that is, phylogenetic) classification because of imperfections in the fossil record. Thus paleontologists must also use a horizontal classification scheme: Related forms that look similar are placed in the same taxonomic categories.

This may sound odd, but it is standard procedure in paleontology. It could be used with the Chinese mole rats as they also underwent parallel evolution. Even George Gaylord Simpson used horizontal classification. He felt that the early horse genus *Merychippus* arose independently from multiple species of the ancestral genus *Parahippus*.[5] Thus a wide consensus existed among paleontologists of the 1940s and 1950s that parallel evolution indeed occurred. The real disagreement was the *meaning* of this pattern of parallelism. Simpson saw it as merely a result of neo-darwinian adaptation to the environment, and as a result of natural selection acting under similar selective pressures. Teilhard saw it as evidence of a deeper, underlying vitalistic force.

Schindewolf eschewed what he felt was a mystical strain of vitalism evident in some of his German paleontological colleagues (not to mention Driesch). Chief among them was Karl Beurlen, the paleontologist who copublished the earliest description of a cloudinid, *Aulophycus*. Beurlen was an active Nazi and felt that evolution was controlled by vitalistic forces. He called these forces the "creative reality of life" and "the will to power." These ideas were consonant with notions of idealistic morphology inherited from the Romantic period of Goethe and the *Naturphilosophen* (natural philosophers), notions accepted by the Nazis.[6]

Schindewolf accepted Beurlen's viewpoint concerning cyclic patterns of evolution known as orthogenesis. The concept of orthogenesis states that organisms have inexorable evolutionary trajectories that drive them in defined evolutionary directions. This much is in common with vitalism. But instead of seeing organisms as being internally driven toward some sort of endpoint (à la Teilhard), Schindewolf saw these evolutionary forces as driving organisms away from some sort of starting point (often toward some unavoidable point of extinction). For Schindewolf, adaptation to the environment played a very minor role. Organisms were internally driven and would follow the same general evolutionary trajectory regardless of when or where they lived. Thus Schindewolf accepted neither the evolutionary ascent toward Teilhard's Omega nor the vitalistic creative power of Beurlen. Nevertheless, Schindewolf shared with these vitalists the fundamental concept that evolution was internally driven, inherent to the organisms themselves, rather than a darwinian affair of adaptation to ambient environment conditions.

For Schindewolf, parallel evolution occurred in related species as a result of the genes they held in common. Similar genes resulted in similar types of mutations. Similarities in possible mutations represented constraints on the amounts and types of variability these organisms could express in an evolutionary sense. So it is no wonder that they evolved in similar ways. Parallel evolution can occur simultaneously in approximately the same place (recall the case of convergence in the Chinese mole rats), but in other instances it happens at different times or in different places. Consider the case Schindewolf cites of the amazing evolutionary convergence of the South African golden mole (*Chrysochloris aurea*) and the South Australian marsupial mole (*Notoryctes typhlops*). Consider also the parallel and independent[7] development of advanced intellectual ability in both Neanderthals (which probably occurred in Europe or the Near East) and *Homo sapiens* (which probably occurred in Africa).

In what he called iterative morphogenesis, Schindewolf described the repetitive copying of evolutionary trends that had appeared in the geological past. He emphasized that these evolutionary repeats were not caused by organisms tracking (in an evolutionary sense) a repeated sequence of environmental changes. Once again, factors internal to the organisms themselves caused the iterated morphologies.

Cases such as these, coupled with Schindewolf's sense that only fairly closely related forms could undergo parallel evolutionary development, led him to largely disregard the importance of the environment in controlling evolutionary change. Thus, similar genotypes may evolve in a limited number of directions, from what Schindewolf would consider to be a shared starting point.

Few (if any) evolutionists today accept the notions of evolutionary heydays and racial senescences of orthogenesis, with all their anti-darwinian implications. Also, most evolutionists grant a large role to the environment in controlling the course of evolution. For example, a highly oxygenated atmosphere will have a decisive influence on the types of evolutionary change possible. Changes can occur in an oxygenated environment that are impossible in an anoxic one.

Intriguing patterns of parallelism, convergence, and iteration still beg for explanation and are not comfortably explained by neo-darwinian model, with its emphasis on "random" mutation as the initiator of evolutionary change. Either evolutionary change is not random (as both Schindewolf and the vitalists would have it) or the environment itself places severe, canalizing constraints on the course of evolution. I suspect that both influences are at play. From the point of view of the cephalo-Ediacaran hypothesis, either the genes shared by both Ediacarans and metazoans or the outside selective pressures that reward "intelligent" life played the dominant role in the development of Ediacaran heads. The task for neovitalists is to determine whether the former (shared genome) or the latter (environmental constraint) is the main motive force behind parallel evolution.

Notes

1. O. H. Schindewolf, *Paläontologie, Entwicklungslehre und Genetik. Kritik und Synthese* (Berlin: Borntraeger, 1936).

2. A. Seilacher, "Otto H. Schindewolf, 7. Juni 1896–10. Juni 1971," *Neues Jarbuch für Geologie und Paläontologie Monatshefte* 2 (1972):69–71.

3. O. H. Schindewolf, *Grundfragen der Paläontologie* (Stuttgart, Germany: Schweizerbart Verlagsbuchhandlung, Erwin Nägele, 1950). This book has been translated into English as *Basic Questions in Paleontology* (Chicago: The University of Chicago Press, 1993). The book was translated by Judith Schaefer, with a forward by Stephen Jay Gould and an afterword by Wolf-Ernst Reif, who also edited this edition. S. Conway Morris ("Wonderfully, Gloriously Wrong," *Trends in Ecology and Evolution* 9 [1994]:407–408) accuses Gould of irony in suggesting that Schindewolf's ideas pose a threat to neo-Darwinism. Conway Morris, himself a firm believer in the reality of convergent evolution, curiously fails to acknowledge that Schindewolf's ideas on convergent and iterative evolution do indeed pose a serious threat to conventional neodarwinism.

4. Haeckel called it "convergence"; E. Haeckel, *Generelle Morphologie der Organismen. Allgemeine Grundzüge der organischen Formen-Wissenschaft mechanisch begründet durch die von Charles Darwin reformierte Descendenz-Theorie. Zweiter Band. Allgemeine Entwickelungsgeschichte der Organismen Kritische Grundzüge der mechanischen Wissenschaft von den entstehenden Formen der Organismen, begründet durch die Descendenz-Theorie* (Berlin: Georg Reimer, 1866).

5. See p. 17 of G. G. Simpson, "The Principles of Classification and a Classification of Mammals," *Bulletin of the American Museum of Natural History* 85 (1945).

6. K. Beurlen, *Die Stammesgeschictlichen Grundlagen der Abstammungslehre* (Jena, Germany: Gustav Fischer Verlag, 1937). Beurlen fled Germany after the war, which led to his encounter with the Brazilian cloudinids.

7. M. Krings, A. Stone, R. W. Schmitz, H. Krainitzki, M. Stoneking, and S. Paabo, "Neanderthal DNA Sequences and the Origin of Modern Humans," *Cell* 90 (1997):19–30.

Appendix

Kingdom Vendobionta Seilacher 1992
Phylum Petalonamae[1] Pflug 1972
Class uncertain
Order uncertain
Family uncertain
Genus *Gehlingia* gen. nov.

Type species: *Gehlingia dibrachida* sp. nov.
Etymology: Named for James G. Gehling.
Diagnosis: A petalonamid with cell families that do not iterate. Two cell families are present and enlarge with growth into paired blade shaped structures, each with multiply bifurcated axes. Bilaterally symmetric between the two paired blades.

Gehlingia dibrachida sp. nov. (figure 2.11)
1988: "unknown frond-like structure," Gehling,[2] p. 308
1994: "enigmatic Ediacaran organism," McMenamin and Mc-Menamin,[3] p. 49

Holotype: South Australian Museum specimen SAM P27927.
Description: A bilaterally symmetric frond-shaped fossil. Each half of the frond is identical and a mirror image to the other half. Each half consists of a swollen axis on the inner edge of the half-frond. This axis bifurcates once, and the bifurcation is directed toward the outer edge of the frond. Numerous tubular structures emanate from the outer edge of the frond axis. These tubules are straight to slightly curved and bifurcate twice before ending abruptly, forming a smooth edge to the frond. A deep groove, as wide as a single axis, separates the paired axes.
Frond was at least 8 cm in length and 3.1 cm in width.
Discussion: This organism was probably most closely related to *Tribrachidium* (figure 1.1). The main difference between the two genera is that whereas *Gehlingia* had two cell families, *Tribrachidium* had three.

Geologic age: Late Lipalian period.
Locality: Ediacara Member of the Rawnsley Quartzite (Pound Subgroup) in the central Flinders Ranges, South Australia.

Notes

1. H. D. Pflug, "Systematik der jung-präkambrischen Petalonamae Pflug 1970," *Paläontologische Zeitschrift* 46 (1972):56–67.

2. J. G. Gehling, "A Cnidarian of Actinian-Grade from the Ediacaran Pound Subgroup, South Australia," *Alcheringa* 12 (1988):299–314.

3. M. A. S. McMenamin and D. L. S. McMenamin, *Hypersea: Life on Land* (New York: Columbia University Press, 1994).

Index

Printed in the USA
CPSIA information can be obtained
at www.ICGtesting.com
JSHW011520221024
72172JS00014B/117